Prediction and Evaluation of Hardened Concrete Strength

Yidong Xu · Jianghong Mao · Weijie Zhuge ·
Xiaoniu Yu · Ping Wu

Prediction and Evaluation of Hardened Concrete Strength

Based on Machine Learning and Mixture Composition

Springer

Yidong Xu
School of Civil Engineering
NingboTech University
Ningbo, China

Jianghong Mao
College of Architecture and Environment
Sichuan University
Chengdu, China

Weijie Zhuge
School of Civil Engineering
Chongqiong Jiaotong University
Chongqing, China

Xiaoniu Yu
School of Materials Science
and Engineering
Southeast University
Nanjing, China

Ping Wu
School of Civil Engineering
NingboTech University
Ningbo, China

ISBN 978-981-96-8236-2   ISBN 978-981-96-8237-9 (eBook)
https://doi.org/10.1007/978-981-96-8237-9

© The Editor(s) (if applicable) and The Author(s) 2026. This book is an open access publication.

Open Access This book is licensed under the terms of the Creative Commons Attribution-NonCommercial-NoDerivatives 4.0 International License (http://creativecommons.org/licenses/by-nc-nd/4.0/), which permits any noncommercial use, sharing, distribution and reproduction in any medium or format, as long as you give appropriate credit to the original author(s) and the source, provide a link to the Creative Commons license and indicate if you modified the licensed material. You do not have permission under this license to share adapted material derived from this book or parts of it.

The images or other third party material in this book are included in the book's Creative Commons license, unless indicated otherwise in a credit line to the material. If material is not included in the book's Creative Commons license and your intended use is not permitted by statutory regulation or exceeds the permitted use, you will need to obtain permission directly from the copyright holder.

This work is subject to copyright. All commercial rights are reserved by the author(s), whether the whole or part of the material is concerned, specifically the rights of translation, reprinting, reuse of illustrations, recitation, broadcasting, reproduction on microfilms or in any other physical way, and transmission or information storage and retrieval, electronic adaptation, computer software, or by similar or dissimilar methodology now known or hereafter developed. Regarding these commercial rights a non-exclusive license has been granted to the publisher.

The use of general descriptive names, registered names, trademarks, service marks, etc. in this publication does not imply, even in the absence of a specific statement, that such names are exempt from the relevant protective laws and regulations and therefore free for general use.

The publisher, the authors and the editors are safe to assume that the advice and information in this book are believed to be true and accurate at the date of publication. Neither the publisher nor the authors or the editors give a warranty, expressed or implied, with respect to the material contained herein or for any errors or omissions that may have been made. The publisher remains neutral with regard to jurisdictional claims in published maps and institutional affiliations.

This Springer imprint is published by the registered company Springer Nature Singapore Pte Ltd.
The registered company address is: 152 Beach Road, #21-01/04 Gateway East, Singapore 189721, Singapore

If disposing of this product, please recycle the paper.

Preface

In this book, the author employs two distinct methods for concrete strength prediction and evaluates their predictive accuracy and applicability: First, by monitoring temperature field development within concrete structures, we established an F-P maturity equation applicable to different temperature ranges based on the Arrhenius equation, investigating how hydration rates influence the strength prediction method of the maturity equation. Second, artificial neural network theory was applied to enhance early-stage concrete strength prediction accuracy, optimizing the ANN model to develop a more precise and broadly applicable predictive framework. The principal research components include:

1. Maturity calculation method based on hydration reaction rate
2. Prediction of compressive strength of hardened concrete using artificial neural network theory
3. Development of an intelligent concrete strength prediction program

This book is completed by Yidong Xu (NingboTech University), Jianghong Mao (Sichuan University), Weijie Zhuge (Chongqing Jiaotong University), Xiaoniu Yu (Southeast University), and Ping Wu (NingboTech University). The authors would like to acknowledge the support from Ningbo Public Welfare Research Project (Grant No. 2024S077), Ningbo Key R&D Program (Grant No. 2024Z287), Ningbo Construction Research Project (Grant No. 2024-23); and National Natural Science Foundation of China (Grant No. 52478281) for publishing this book. The authors would also like to acknowledge the support from Zhou Ronghan from Sichuan University for her contributions to textual translation and graphic proofreading of this book. In addition, the author extends sincere gratitude to Cheng Yanhua from

NingboTech University, for her meticulous proofreading and valuable suggestions, which have greatly enhanced the quality of this book.

Ningbo, China Yidong Xu
Chengdu, China Jianghong Mao
Chongqing, China Weijie Zhuge
Nanjing, China Xiaoniu Yu
Ningbo, China Ping Wu

Contents

1 **Introduction** 1
   1.1 Research Background and Basis 1
   1.2 Review of Research Status 2
      1.2.1 Prediction of Concrete Strength Based on Maturity Equation 2
      1.2.2 Prediction of Concrete Strength Based on Artificial Neural Networks 7
      1.2.3 Development of an Intelligent Prediction Programme for Concrete Strength 9
   1.3 Research Idea 10
      1.3.1 Research Objective 10
      1.3.2 Research Content 11
      1.3.3 Technical Route 11
   References 14

2 **Raw Materials and Experimental Method** 17
   2.1 Raw Materials 17
   2.2 Experimental Apparatus 18
   2.3 Mix Design 18
   2.4 Experimental Content and Methodology 20
      2.4.1 Experimental Preparation 20
      2.4.2 Experimental Methods and Procedures 21
   2.5 Summary of the Chapter 21
   References 22

3 **Establishment of Maturity Equations for Different Temperature Intervals** 23
   3.1 Quantitative Study of the Effect of Temperature Change on the Rate Constants of Cement Hydration Reaction 23

3.2 F-P Maturity Model Construction Based on Specific
Temperature Intervals 30
    3.2.1 Determination of Apparent Activation Energy
in the Positive and Low Positive Temperature Range 30
    3.2.2 Relationship Between Equivalent Age and Compressive
Strength 32
3.3 Analysis of the Prediction Accuracy of the F-P Maturity
Equation 34
3.4 Improving the F-P Maturity Equation 35
3.5 Summary of the Chapter 38
References 39

4 Concrete Strength Prediction Based on Artificial Neural Networks ... 41
4.1 Establishment of BP Neural Network Model 41
    4.1.1 BP Neural Network Models 41
    4.1.2 BP Neural Network Design Process 42
    4.1.3 BP Neural Network Input and Output Layers 44
    4.1.4 Results and Analyses 45
4.2 BP Neural Network Based on Particle Swarm Optimization
Algorithm 47
    4.2.1 Overview of Particle Swarm Optimization Algorithm 47
    4.2.2 PSO-BP Modelling 48
    4.2.3 Analysis of Operational Results 49
4.3 BP Neural Network Based on Ant Colony Optimization
Algorithm 53
    4.3.1 Overview of Ant Colony Optimization Algorithm 53
    4.3.2 ACO-BP Modelling 55
    4.3.3 Analysis of Operational Results 56
4.4 Comparison of Model Prediction Performance 58
4.5 Summary of the Chapter 60
References 63

5 Development of Intelligent Concrete Strength Prediction Program ... 65
5.1 Analysis and Implementation of Intelligent Concrete Strength
Prediction Program 65
    5.1.1 Theoretical Foundations of MATLAB Program
Development 65
    5.1.2 Intelligent Program Function Module Design 66
    5.1.3 Intelligent Program Design Process 67
5.2 Intelligent Concrete Strength Prediction Program Design 68
5.3 Concrete Strength Intelligent Prediction Program Performance
Measurement 70
5.4 Summary of the Chapter 70

6	**Conclusions and Foresight**	75
	6.1 Conclusions of the Study	75
	6.2 Innovation Points	76
	6.3 Future Prospects	76

Appendix: Summary of Hardened Concrete Compressive Test Data Used in This Study 79

Chapter 1
Introduction

1.1 Research Background and Basis

Reinforced concrete structures dominate the modern construction engineering field due to their stability and durability. The strength of concrete is one of the most critical factors in ensuring the safety of building structures. According to the Code for Quality Acceptance of Concrete Structure Construction (GB50204-2015), concrete strength grade acceptance is the primary consideration during the quality inspection of concrete projects. This code clearly stipulates that the assessment of concrete strength must strictly comply with design standards to ensure the long-term safety and reliability of the structure [1, 2]. Currently, there is a lack of effective methods for predicting and evaluating early concrete strength. The performance of early concrete is influenced by various factors, including but not limited to material composition, environmental conditions, and construction methods, making accurate prediction of its strength development a complex task. Therefore, there is an urgent need in the field of building materials science to explore and develop novel technologies, methods, and tools aimed at improving the accuracy and reliability of early concrete strength prediction.

In practical engineering applications, the evaluation of the strength of concrete structures mainly relies on several classical methods, including the compressive strength test of specimens subjected to the same condition of curing, the rebound method using rebound gauges, or the combined ultrasonic rebound method in combination with ultrasonic testing, and the core drilling and sampling method. Each of these methods has different advantages and limitations, but all of them are dedicated to accurately assessing the load-bearing capacity of concrete structures to ensure the structural safety and lasting stability of buildings. Exploring more efficient and accurate prediction techniques is a key way to improve building safety performance and economic efficiency. However, these testing and evaluation methods are mostly used for the presumption of concrete strength of the solid structure after the completion of pouring, and the evaluation of concrete strength in China is generally based on the

Standards for Evaluation of Concrete Compressive Strength (GB/T 50,107-2010) for the determination of 28-day strength of standard specimens, and it is usually only after 28 days that it can be judged whether the structural strength meets the design requirements, which belongs to the post-control measures, and it lacks of timeliness and representativeness. There are obvious deficiencies, and once the strength is found to be unqualified, it will have to cost a lot of money to make up reinforcement, or even lead to the scrapping of the project, resulting in significant economic losses [3].

Nowadays, the rapid development of artificial intelligence technology has established a close connection with the civil engineering industry, whether through artificial neural networks or deep learning technologies. AI can provide innovative driving forces for research and application in various fields, deeply integrating with concrete strength prediction to provide safety assurance for actual engineering projects [4]. The development of intelligent prediction programs for compressive strength is an important future direction for the civil engineering industry. Therefore, it is necessary to conduct research on early assessment of concrete quality issues and prediction of concrete strength at construction sites, gradually incorporating intelligent theories into the prediction of concrete compressive strength. This study will utilize two different prediction methods based on the composition and characteristics of the concrete mixture, establish the inherent relationship between them, use evaluation indicators to determine accuracy, improve the prediction and evaluation system for hardened concrete strength, and create an intelligent prediction program. The main objective of this book is to develop an intelligent prediction software for rapid and accurate prediction of concrete compressive strength, improving quality control and optimizing construction processes by enhancing existing prediction methods.

1.2 Review of Research Status

1.2.1 Prediction of Concrete Strength Based on Maturity Equation

(1) Relationship between maturity and concrete strength

The maturity method is based on the fundamental theory that there is a certain relationship between the maturity index and the concrete strength. Even with different curing temperatures and times, the same maturity level corresponds to the same strength [5]. Compared to previous studies, Saul's concept of temperature–time factor, which represents the 'maturity' of concrete, is widely used. However, this method simplifies the impact of temperature on strength as a linear relationship, limiting its application range [6, 7]. Hansen and Pedersen [8] proposed a new maturity function based on the Arrhenius function to represent the impact of temperature, overcoming the limitations of Saul's linear function. Currently, two widely accepted maturity models are primarily used in calculating concrete maturity. The first is the

1.2 Review of Research Status

traditional Nurse-Saul maturity model, also known as the N-S temperature–time factor equation. The second is the F-P maturity equation named after Hansen and Pedersen. Both equations provide unique computational framework and theoretical foundation for the relationship between concrete strength and maturity.

Concrete strength development is closely related to the temperature and time it experiences. Saul proposed that the product of temperature and time during concrete hardening could effectively represent strength development. He named this 'maturity', expressed as the product of temperature–time.

Rastrup [9] and Cao et al. [10] introduced the concept of equivalent age te, which is the ratio of the time experienced at the actual curing temperature of the concrete to the time experienced at temperature Tr, while ensuring the same maturity value is reached.

Ignoring the impact of cement hydration on the evaluation of concrete performance can render the obtained maturity meaningless. Therefore, the equivalent age-maturity function developed empirically by Hansen and Pedersen can more accurately address this issue. This function is based on the Arrhenius equation, and its mathematical expression is as follows:

$$K(T) = e^{\frac{-E_a}{RT}} \tag{1.1}$$

$K(T)$—reaction rate constant (d^{-1});
R—gas constant;
E_a—apparent activation energy (kJ/mol);
T—temperature (°C).

Hansen and Pedersen used the equation above to improve Saul's maturity formula by defining k as the cement hydration rate, and the mathematical expression of the improved equation is as follows:

$$M_a = \sum_0^t k \, \Delta t = A \sum_0^t e^{(-\frac{E_a}{RT})} \, \Delta t \tag{1.2}$$

where A is coefficient that need to be determined experimentally, Δt is the time interval during the hardening process of concrete.

Here the physical meaning of the value of apparent activation energy E_a can be understood as the minimum energy required for the cement hydration. Obviously, the higher the temperature, the smaller the energy required, and vice versa, the lower the temperature, the larger the energy required for cement hydration [7, 11]. Therefore, the current method of determining the value of E_a is given in the following equation:

$$E_a = 33.5 + 1.47(20 - T) kJ/mol \, , T < 20°C$$

$$E_a = 33.5 kJ/mol \, , T \geq 20°C \tag{1.3}$$

Fig. 1.1 Relationship between apparent activation energy and cement hydration time [14]

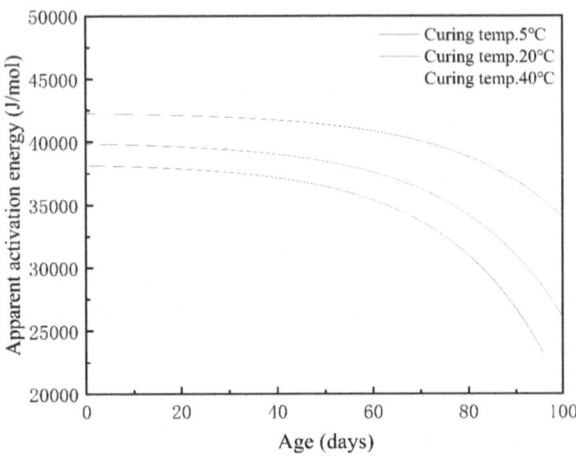

Kjellsen et al. [12] and Hemstad et al. [13] consider the influence of the 'hysteresis' effect on the late strength growth of concrete and propose that the apparent activation energy E_a is a function of hydration degree (Fig. 1.1).

Yikici and Chen [15] and Nicholas et al. [16] consistently agreed that the Arrhenius equation is superior to the linear equation in representing the effect of temperature on the hydration rate, and in engineering practice, the original equation is often modified using a reference temperature to simplify the equivalent age expression. The mathematical expression is as follows:

$$t_e = \sum e^{B(T-T_r)} \Delta t \qquad (1.4)$$

B—temperature sensitivity factor;
T—average concrete temperature (°C) at time Δt;
T_r—reference temperature (°C).

These models were proposed to explore the early performance of concrete. However, as mentioned above, due to differences in the hydration process, curing time, and strength development, many experimental results show that considering maturity from the perspective of cement hydration is a more accurate method. For concrete with the same cement components, as long as their hydration degree is the same the concrete strength should be the same as well.

Equally important to the maturity index is the strength prediction model, because no matter how accurate the estimation of the maturity index is, it is meaningless if the strength prediction is inaccurate. The best fitting smooth curve can be drawn through data, or regression analysis can be used to determine the best fitting curve for the appropriate strength–maturity relationship [16, 17].

Plowman [18, 19] proposed a relationship between strength and maturity with the following mathematical expression:

1.2 Review of Research Status

$$S = a + b \log(M) \tag{1.5}$$

S—Concrete strength (MPa);
a, b—constants;
M—Maturity (°C·d).

This equation, as a popular current predictive model, is simple in form and convenient to use, but it also has its flaws. Briefly, it accurately predicts the strength of intermediate maturity values by drawing a straight line, but it does not describe the maturity indices at low or high values well [20].

Reichard [21] proposed the following nonlinear regression formula by analysing the relationship between compressive strength test results and maturity:

$$S = \frac{K}{1 + Ka[\log(M - 30)]^b} \tag{1.6}$$

K, a, b—numerical constants;
M—maturity (°C·d);
30—maturity when the effective strength value is 0;
S—concrete strength (MPa).

Hansen and Pedersen [22, 23] proposed the following exponential equation to represent the strength development of concrete:

$$S = S_u e^{-[\tau/M]^\alpha} \tag{1.7}$$

S_u—ultimate strength (MPa);
τ—time constant;
α—shape parameter.

The Gompertz [8, 22] curve equation is also widely used today to predict the compressive strength of concrete. The curve features a rapid rise, deceleration, and then approaches a horizontal state, and it is usually used to describe the development cycle of things from germination, growth to saturation, which closely resembles concrete strength development. The inflection point of the curve is at S_u/e and the curve does not exhibit bilateral symmetry. Its mathematical expression is as follows:

$$S = S_u \cdot e^{(-a \cdot e^{-b} \cdot \log M)} \tag{1.8}$$

S—concrete strength (MPa);
S_u—ultimate strength (MPa);
a, b—numerical parameters;
M—maturity (°C·d).

Based on the influence of different cement varieties on the correlation between maturity and strength, Li et al. [24] proposed the strength equations for the most widely used ordinary Portland cement and Portland slag cement concrete with the following mathematical expressions:

$$S = 10.76M^{0.331}(0 \leq M \leq 840°C \cdot d) \tag{1.9}$$

$$S = 2.32M^{0.559}(0 \leq M \leq 840°C \cdot d) \tag{1.10}$$

As mentioned above, many models relating maturity and concrete strength have been proposed today, and these models were introduced before the widespread use of artificial neural networks and other machine learning technologies, with both accuracy and theoretical basis generally well-established.

(2) Application research of maturity equation

In the application of maturity, Galobardes et al. [26] used the N-S maturity equation to predict the compressive strength of sprayed concrete. They prepared 24 different proportions of concrete in the laboratory and recorded their temperature history and compressive strength data. The results showed that the N-S maturity equation could accurately predict the strength development pattern of sprayed concrete under specific conditions. The study also revealed that the maturity curve is influenced by multiple factors such as cement type, strength grade, type and amount of accelerator.

Han and Han [25] systematically investigated the effects of retarders on concrete setting time at different curing temperatures using the F-P maturity equation, and assessed its prediction accuracy. They observed that after converting concrete setting time to equivalent age at different temperatures, the value showed high stability, almost constant. The predictions were highly consistent with experimental data, confirming that the maturity method can effectively predict the setting time of concrete with retarders. The experiment also revealed a significant change in the apparent activation energy (E_a) of initial and final setting times with varying amounts of retarder.

Zhou et al. [27] conducted systematic curing experiments on concrete with ultra-fine fly ash under strictly controlled conditions (8 ± 3 °C, RH about 80%) to investigate the effect of curing time on strength development of concrete after steam curing. The results indicated that when the curing time was extended beyond 4 h, reaching a critical point at 6 h, the concrete exhibited peak compressive strength at 28 days, showing optimal mechanical property.

Jin et al. [5] used the maturity equation to predict the early compressive strength of vinyl ester polymer concrete. The results showed a rapid increase in strength before 24 h of curing, and a significant slowdown in strength growth after 72 h. As the content of methyl methacrylate increased, phase separation phenomenon had an increasingly negative impact on concrete strength, leading to a reduction in strength.

Guo et al. [28] conducted experimental and modelling studies on the effects of setting temperature (T) and relative humidity (RH) on the early compressive strength development of cement mortar, proposing a modified maturity function that considers the impact of T and RH on concrete mechanical properties. The compressive strength of the mortar decreased with lower curing relative humidity, with this reduction becoming more pronounced in later stages of curing. Under the same relative humidity curing conditions, samples achieved higher early strength at higher temperatures [9].

In summary, both the N-S and F-P maturity equations point to the need to obtain the apparent activation energy (E_a) to determine the reaction rate constant, which is ultimately converted to equivalent age to predict concrete strength.

1.2.2 Prediction of Concrete Strength Based on Artificial Neural Networks

(1) Working principle of artificial neural networks

Artificial neural networks consist of many interconnected neurons, each capable of independently processing information. McCulloch and Pitts [29] abstracted the propagation process of biological neural networks into the still-used M-P neuron model. Artificial neural networks, where multiple neurons are connected in a certain hierarchical structure [30], serve as a mathematical model that simulates the complex structure of biological neurons and have self-learning capabilities [31]. Thus, the function of a neuron is to calculate the inner product of an input vector and a weight vector, then apply a nonlinear transfer function to produce a scalar output. Nowadays, artificial neural networks can already simulate the human brain's thought processes and make corresponding predictions by continuously training to learn the relationship between inputs and outputs, thereby reducing errors to optimize predictions. Experimental data is used to construct the neural network and 'learn' the relationships among a large number of highly interconnected processing units to approximate the results of other experiments.

Artificial neural networks transform the input space to the output space into a high-dimensional nonlinear mapping, which is particularly suitable for predicting the compressive strength of concrete. This is because in the algorithm structure, the neurons in the input layer, hidden layer, and output layer mutually constrain each other, as shown in Fig. 1.2 [32]. Neural networks not only have self-learning capabilities but also possess associative memory functions and the ability to rapidly find optimal solutions. This is why, with sufficient datasets and defined input variables, they perform exceptionally well in predicting the compressive strength of concrete.

(2) Prediction of concrete strength by artificial neural networks

Chopra et al. [33] used artificial neural networks (ANN) and multiple linear regression to compare the compressive strength of concrete at 7 and 28 days. They found that the two key indicators reflecting prediction error could be reduced by more than 10%, indicating that AI systems based on ANN models can achieve a level of prediction accuracy that is difficult with traditional statistical methods. Some evolutionary methods in artificial neural networks have optimized hyper-parameter settings, model training efficiency, and generalization capabilities, giving them a competitive edge

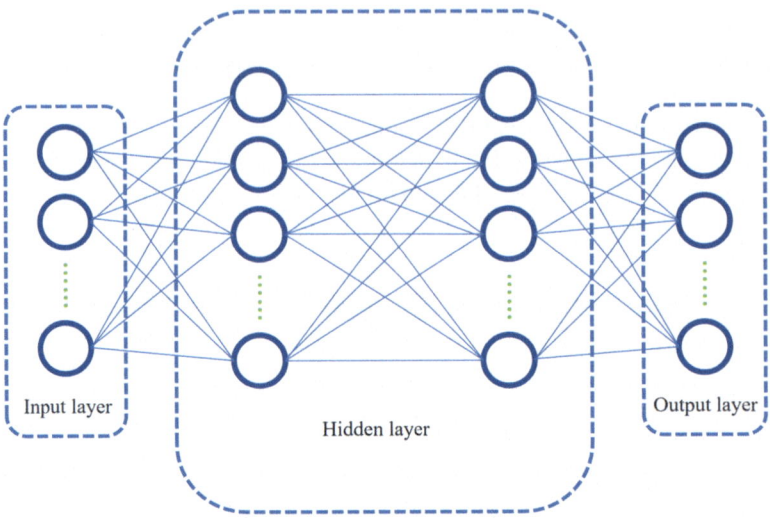

Fig. 1.2 ANN composition structure

in predicting concrete strength. Rafiei et al. [34] combined unsupervised learning methods with ANNs to extract features from datasets, allowing for quicker adaptation to data evolution and improved prediction accuracy while reducing computational costs. Dutta et al. [35] Improved ANNs with fuzzy logic, creating an adaptive neuro-fuzzy inference system (ANFIS) that showed good applicability and portability in predicting concrete strength. However, the variety of available models necessitates identifying the most suitable ones for hardened concrete strength prediction.

Asteris et al. [36] applied Artificial Neural Networks (ANN) to predict the 28-day compressive strength of self-compacting concrete with admixtures. Mohammed et al. [37] adopted nonlinear regression models (NLR) and ANN techniques for qualitative analysis of the strength of nano-clay modified cement paste. Based on static analysis, both simulation techniques predicted the compressive strength well, but the NLR model based on experimental datasets was found to be the most reliable for predicting cement compressive strength, outperforming the ANN model.

In summary, artificial neural networks are more adaptable to the diversity of concrete and the complexity of matching components than traditional statistical methods and non-destructive testing techniques, making ANN predictions of concrete strength of significant reference value.

(3) Artificial neural network strength prediction algorithm and its accuracy

After setting up an artificial neural network, how to improve its prediction accuracy becomes a key focus. In terms of the dataset, one can expand the dataset or increase its diversity and randomness. For data pre-processing, enhancing the data's generalization capabilities, such as through data centerization and data standardization, is beneficial. During the training process, adjusting the batch-size is crucial as it determines the time required for each epoch and the smoothness of gradient updates between iterations. Shuffling the order of data in the dataset can prevent sequence biases in training. Adjusting the learning rate is essentially an optimization process that gradually moves towards the optimal solution. Weight decay can also help reduce model overfitting to some extent.

Sada and Ikpeseni [38] demonstrated the application of the Levenberg–Marquardt (LM) and Scaled Conjugate Gradient (SCG) algorithms in predicting the metal removal rate (MRR) and surface roughness (Ra) in low-carbon steel pipe processing, which belongs to updating weights to improve accuracy. Chen et al. [39] incorporated experiential knowledge from concrete engineers' on-site batching, known as interpretable features (strength factors), into the neural network model to enhance prediction accuracy and meet actual construction needs. The article proposes a BPNN prediction model optimized by an improved sparrow search algorithm (ISSA) and random forest (RF) to enhance the generalization and prediction accuracy of BPNN for concrete compressive strength. Mohammadi Golafshani et al. [40] aimed to accurately estimate the chloride diffusion in reinforced concrete to reliably predict its service life. They used metaheuristic algorithms to balance exploration and exploitation, overcoming the overfitting issue present in artificial neural networks trained with classical optimization algorithms. Li et al. [41] and Luo et al. [42] utilized finite element modelling (FEM) and artificial neural networks (ANN) to evaluate the direct tensile strength of BFRC, employing a tan-sigmoid transfer function for the input-hidden and hidden-output layers, achieving the most accurate predictions. Each optimization method varies in its effectiveness at improving prediction accuracy for different types of neural network models, and finding the optimal optimization algorithm remains a topic for exploration and in-depth study.

1.2.3 Development of an Intelligent Prediction Programme for Concrete Strength

For the safety inspection of existing engineering projects and the development of rapid on-site inspection work, traditional methods for predicting concrete strength are concrete strength is inefficient and lacks timeliness and real-time capabilities. Consequently, the development of artificial intelligence technology in recent years has significantly advanced the civil engineering industry. There is a pressing need for the development of intelligent concrete strength prediction programs to enhance rapid on-site testing in project management departments.

Both the maturity equation and artificial neural network methods mentioned in Sects. 1.2.1 and 1.2.2 for predicting concrete strength, as well as other methods, have their limitations and are not universally applicable to all types of engineering projects. Due to the efficient computational capability of MATLAB software, which is increasingly recognized by scholars, these methods can be integrated using MATLAB to create an intelligent concrete strength prediction program.

Programmes on computers need to be run using a standalone programme runtime in MATLAB, especially when developing and deploying complex computational models on Windows, the mainstream operating system. By utilizing MATLAB, developers can integrate the concrete strength prediction model directly into the code. This significantly reduces the development and maintenance costs of the program and leverages the powerful self-learning capabilities of artificial neural networks (ANN), providing an automated and efficient way for database updates and model iterations. With the rapid development of information technology, the diversity of basic concrete parameters and engineering environments requires the model to handle data of different dimensions and adapt to various complex environmental conditions. The development of an intelligent compressive strength prediction program offers the following advantages:

(1) As the intersection of civil engineering and computer science deepens, an increasing number of students and professionals are turning to MATLAB to tackle engineering challenges. This reflects not only advancements in technical education but also a revolution in on-site construction practices. Developing efficient and intelligent MATLAB-based programs for rapid inspection and analysis of construction sites could significantly enhance project efficiency and safety.
(2) Intelligent programs developed based on MATLAB can be easily deployed on computer systems at construction sites. Users simply need to install the corresponding MATLAB runtime environment on their computers to start and run the program. This program features a clear and straightforward interface, allowing users to quickly obtain predictive results by entering key parameters, greatly enhancing user experience and effectively addressing practical issues encountered on-site.
(3) In the context of rapidly evolving information technology, the literature on concrete property is growing, and related databases are increasingly open to the public. This provides a rich resource base for the continuous updating and improvement of intelligent programs. Compared to traditional prediction methods, MATLAB-based intelligent programs can not only update databases in real time but also quickly adapt to changes in engineering needs, ensuring timely and accurate safety monitoring and assessment at construction sites, thus accelerating the initiation and progress of engineering projects.

1.3 Research Idea

1.3.1 Research Objective

(1) To monitor the development of the internal temperature field in concrete structures on-site, and establish the F-P maturity equation for different temperature ranges based on the Arrhenius formula. This involves systematically studying the accuracy and applicability of strength prediction methods based on the

1.3 Research Idea

hydration rate maturity equation, providing a reference for further research on concrete strength prediction and evaluation systems.

(2) To further expand experimental research and verify the reliability of test results. Based on artificial neural network (ANN) theory, propose using ANNs to predict early-age concrete strength. Optimize the ANN prediction model to achieve higher prediction accuracy and broader applicability.

(3) To verify the applicability and convenience of different prediction methods in various environmental regions, two methods for predicting early concrete strength are compared and analysed to find the correlation between them. Based on maturity theory and the artificial neural network model, developing an intelligent concrete strength prediction program using MATLAB facilitates rapid prediction and evaluation of strength based on on-site measured parameters.

1.3.2 Research Content

(1) Maturity calculation method based on hydration reaction rate: Establish the F-P maturity equation to predict strength using the determined reference temperature value and apparent activation energy.

(2) Predicting the compressive strength of hardened concrete based on artificial neural network theory: Construct a compressive strength prediction model using the typical BP neural network. Due to the limitations of the BP neural network, it is necessary to optimize the prediction model, proposing different types of neural networks optimized by particle swarm algorithms and ant colony algorithms, developing a neural network model with higher prediction accuracy.

(3) To further improve the research on the strength prediction and evaluation system, assist with the method of predicting concrete strength using the technically refined maturity equation to ensure the accuracy and reliability of predictions. Use indicators such as root mean square error (RMSE) to comprehensively analyse different prediction methods and determine the correlation between the two prediction methods to facilitate rapid prediction at the construction site.

1.3.3 Technical Route

(1) Maturity calculation method based on hydration reaction rate: Establish a nonlinear relationship between maturity and concrete strength. After determining the reference temperature value at the construction site, calculate the corresponding apparent activation energy, convert it to equivalent age, and then establish the F-P maturity equation for strength prediction.

(2) Use a typical BP neural network to build a compressive strength prediction model, with cement, water, coarse aggregate, fine aggregate, and fly ash as input parameters, and concrete strength as the output parameter. Due to the poor

generalization performance of the BP neural network, the calculation accuracy is difficult to guarantee. It is necessary to optimize the prediction model, proposing different types of neural networks optimized by particle swarm optimization (PSO-BP) and ant colony optimization (ACO-BP) to develop a neural network model with higher prediction accuracy.

(3) With the help of the technically refined maturity equation method to predict concrete strength, we can ensure the accuracy and reliability of concrete strength predictions. Indicators such as the root mean square error (RMSE) are used to comprehensively analyse different prediction methods, determine when and where to apply each method for strength prediction, establish correlations between the methods, and develop intelligent computing programs for rapid on-site prediction at construction sites. The technical route adopted in this topic is shown in Fig. 1.3.

1.3 Research Idea

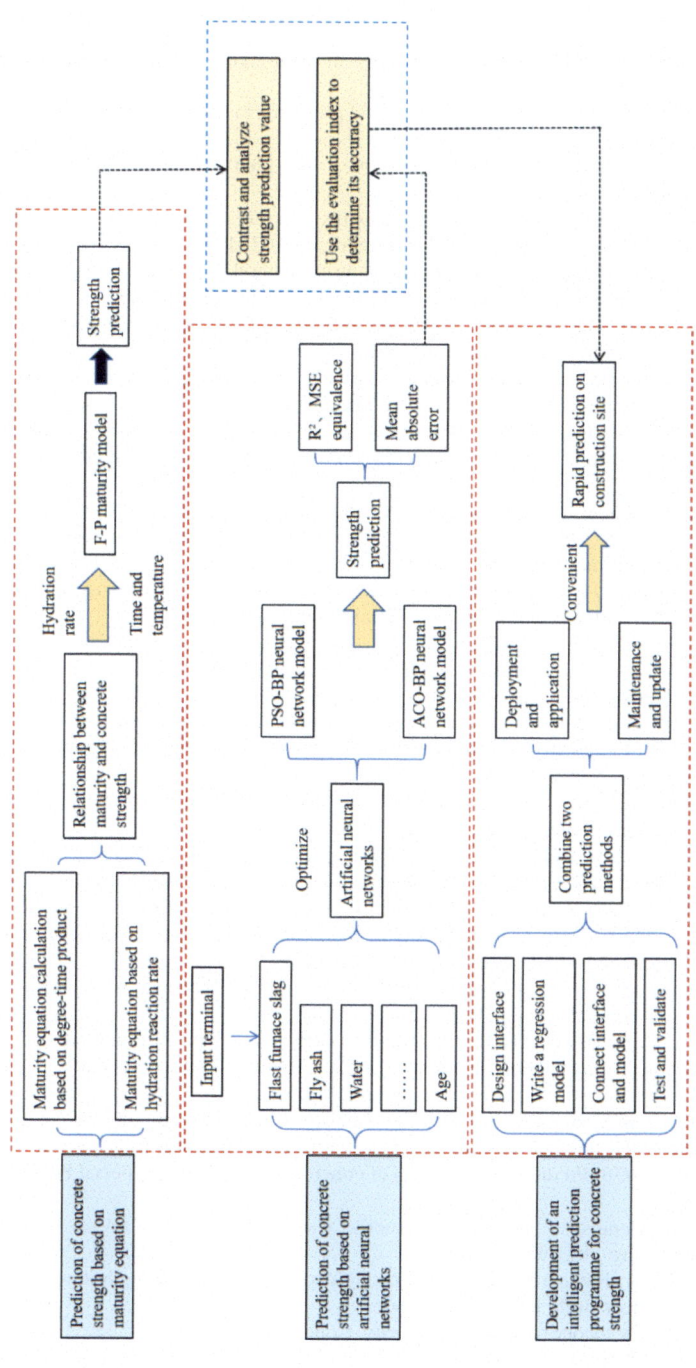

Fig. 1.3 Proposed technical route

References

1. Luo J, Bai GS (2023) Evaluation of uncertainty in testing concrete compressive strength. Zhuanwa 49(6):44–45
2. Chang XD, Yang YX (2012) Discussion on several problems in testing techniques of concrete strength. Construct Qual 30(10):68–70 (2012)
3. Standard for test method of mechanical properties on odinary concrete GB/T50081–2002 (2007)
4. Luo DM, Li F, Niu DT (2024) Research progress on durability diagnosis of concrete structuresbased on artificial intelligence. J Build Struct 45(02):1–13
5. Jin NJ, Yeon K, Min S, et al (2017) Using the maturity method in predicting the compressive strength of vinyl ester polymer concrete at an early age. Adv Mater Sci Eng 1–12
6. Yu GN, Sun XL (2012) Prediction of early strength of cement concrete pavement by means of maturity method. Technol. Highway Transp (5):15–18
7. Ma Y, Ma L, Pan BL (2019) Research on predicting early strength of concrete based on maturity method. Technol. Highway Trans (Appl Technol Edn) 5:1–4
8. Sun B, Noguchi T, Cai G et al (2021) Prediction of early compressive strength of mortars at different curing temperature and relative humidity by a modified maturity method. Struct Concr 22(S1):732–744
9. Rastrup E (1954) Heat of hydration in concrete. Mag Concr Res 17(6):79–92
10. Cao Y, Wang RW, Ou DF (2023) Research on estimation of concrete strength for winter construction based on maturity theory. Highway 68(09):367–373
11. Liao YS, Gui Y, Shen Q et al (2018) Determination of Apparent activation energy of hydration reaction of calcium sulfoaluminate cement. J Build Mater 21(6):864–870
12. Kjellsen KO, Detwiler RJ, Gjørv OE (1991) Development of microstructures in plain cement pastes hydrated at different temperatures. Cement Concr Res 21(1):179–189
13. Hemstad P, Lothenbach B, Kjellsen KO et al (2024) The effect of varying cement replacement level on alkali metal distribution in cement pastes. Cem Concr Compos 146:105344
14. Kim JK, Han SH, Lee KM (2001) Estimation of compressive strength by a new apparent activation energy function. Cem Concr Res 31(2):217–225
15. Yikici TA, Chen HR (2015) Use of maturity method to estimate compressive strength of mass concrete. Constr Build Mater 95:802–812
16. Nicholas J. Carino M A, And H. S. Lew F A. The maturity method: from theory to application. American Society of Civil Engineers (ASCE), Washington, DC, United states
17. Duan Y, Wang Q, Yang Z, et al (2022) Research on the effect of steam curing temperature and duration on the strength of manufactured sand concrete and strength estimation model considering thermal damage. Construct Build Mater 315:125531
18. Kapoor H, Kaufman JC (1956) Conclusion[M]. In: Kapoor H, Kaufman JC (eds) Creativity and morality. Academic Press, pp 303–304
19. Quan L, Tian B, Feng DC, et al (2012) Research on early strength prediction model of cement concrete based on maturity theory. Highway Traf Sci Technol (Applied Technology Edition) 8(02):35–39
20. Jin NJ, Yeon K, Min S, et al (2017) Using the maturity method in predicting the compressive strength of vinyl ester polymer concrete at an early age. Adv Mater Sci Eng 1–12
21. Reichard HSLA (1978) Prediction of strength of concrete from maturity. Special Publ 56:229–248
22. Helal J, Sofi M, Mendis P (2015) Non-destructive testing of concrete: a review of methods. Electron J Struct Eng 14(1):97–105
23. Sofi M, Mendis PA, Baweja D (2012) Estimating early-age in situ strength development of concrete slabs. Constr Build Mater 29:659–666
24. Li FQ (2002) A new theory of concrete maturity. Concrete 10:12–14
25. Han M, Han C (2010) Use of maturity methods to estimate the setting time of concrete containing super retarding agent. Cem Concr Compos 32(2):164–172

26. Galobardes I, Cavalaro SH, Goodier CI et al (2015) Maturity method to predict the evolution of the properties of sprayed concrete. Constr Build Mater 79:357–369
27. Zhou WX, Xie YJ, Sun LJ (2003) The influence of steam curing regime on the strength of ultra-fine fly ash concrete. Concrete 6:35–37
28. Guo XY, Fang KH, Leng FG, et al (2005) Experimental study on effect of environment conditionson properties of HSC. J Yangtze River Sci Res Inst 22(2):57–59
29. Zhang S, Yu Y, Wang H (2015) Mittag-Leffler stability of fractional-order Hopfield neural networks. Nonlinear Anal Hybrid Syst 16:104–121
30. Mu QH, Wang YT, Lv YS et al (2023) In celebration of McCulloch-Pitts ANN model's 80thanniversary: its origin, principle, and influence. Chinese J Intell Sci Technol 5(02):133–142
31. Lu QS, Liu SQ, Liu F et al (2008) Research on dynamics and functions of biological neural network systems. Adv Mech 06:766–793
32. Jia M (2017) Computation method for unstable manifold of high-dimensional nonlinear mapping systems. Vibrat Shock 36(17):262–266
33. Chopra P, Sharma RK, Kumar M et al (2018) Comparison of machine learning techniques for the prediction of compressive strength of concrete. Adv Civil Eng 2018:1–9
34. Rafiei MH, Adeli H (2018) A novel unsupervised deep learning model for global and local health condition assessment of structures. Eng Struct 156:598–607
35. Dutta P, Kumar A (2018) Application of an ANFIS model to optimize the liquid flow rate of a process control system. Chem Eng Trans 71
36. Asteris PGKKGD (2016) Prediction of self-compacting concrete strength using artificial neural networks[J]. Eur J Environ Civ Eng 20(1):102–122
37. Mohammed A, Rafiq S, Mahmood W et al (2021) Artificial Neural Network and NLR techniques to predict the rheological properties and compression strength of cement past modified with nanoclay. Ain Shams Eng J 12(2):1313–1328
38. Sada SO, Ikpeseni SC (2021) Evaluation of ANN and ANFIS modelling ability in the prediction of AISI 1050 steel machining performance. Heliyon 7(2):6136
39. Chen G, Zhu D, Wang X, et al (2022) Prediction of concrete compressive strength based on the BP neural network optimized by random forest and ISSA. J Funct Spaces 1–20
40. Mohammadi Golafshani E, Kashani A, Kim T, et al (2022) Concrete chloride diffusion modelling using marine creatures-based metaheuristic artificial intelligence. J Clean Prod 374:134021
41. Li H, Lin J, Zhao D et al (2022) A BFRC compressive strength prediction method via kernel extreme learning machine-genetic algorithm. Constr Build Mater 344:128076
42. Luo B, Huang WJ, Yang S (2006) Improved algorithm of BP neural network based on parameters adjustment of Tan-Sigmoid transfer function. J Chongqing University (Natural Sciences Edition) (01):150–153

Open Access This chapter is licensed under the terms of the Creative Commons Attribution-NonCommercial-NoDerivatives 4.0 International License (http://creativecommons.org/licenses/by-nc-nd/4.0/), which permits any noncommercial use, sharing, distribution and reproduction in any medium or format, as long as you give appropriate credit to the original author(s) and the source, provide a link to the Creative Commons license and indicate if you modified the licensed material. You do not have permission under this license to share adapted material derived from this chapter or parts of it.

The images or other third party material in this chapter are included in the chapter's Creative Commons license, unless indicated otherwise in a credit line to the material. If material is not included in the chapter's Creative Commons license and your intended use is not permitted by statutory regulation or exceeds the permitted use, you will need to obtain permission directly from the copyright holder.

Chapter 2
Raw Materials and Experimental Method

This chapter provides a detailed description of the various raw materials used in the experiment and lists their specific types and properties. Additionally, it also elaborates on the design methodology and experimental procedures adopted in this experiment, as well as the related experimental equipment employed. A comprehensive explanation of the functions and operation of each piece of equipment will be given to ensure that readers fully understand the implementation process of the experiment and its scientific basis. Through this presentation, the chapter aims to provide readers with a clear and comprehensive overview of the experimental process, thereby facilitating a better understanding of how the experimental results and conclusions are derived.

2.1 Raw Materials

(1) Cement: The cement used in this experiment is P-O42.5 normal Portland cement produced by Ningbo Conch Cement Co. Its chemical composition is shown in Table 2.1 and its basic properties are shown in Table 2.2.

(2) Sand: The sand selected for the experiment is medium sand. According to Standard for technical requirements and test method of sand and crushed stone (or gravel) for ordinary concrete (JGJ 52–2006), its apparent density is 2590 kg/m^3, and its bulk density is 1440 kg/m^3.

(3) Crushed stone: The experiment selects 5-20 mm continuously graded crushed stone. According to Standard for technical requirements and test method of sand and crushed stone (or gravel) for ordinary concrete (JGJ 52–2006), its apparent density is 2750 kg/m^3, and its bulk density is 1580 kg/m^3.

(4) Fly ash:

The Class I fly ash selected for this experiment is sourced from Ningbo, Zhejiang, with a water demand ratio of 95% and a loss on ignition(LOI)of 1.39%

(5) Water reducer:

Table 2.1 Chemical composition of Conch Brand P-O42.5 cement

Al_2O_3 (%)	SiO_2(%)	CaO (%)	Fe_2O_3 (%)	MgO(%)	SO_3 (%)	Loss on ignition (%)
5.98	21.24	60.98	4.35	3.56	2.49	1.4

Table 2.2 Basic performance indicators for cement

Water content for normal consistency (%)	Initial setting time (min)	Final setting time (min)	Specific surface area (m^2kg^{-1})	Soundness
25.9	139	198	376	Qualified

The water reducer used in this experiment is PCA-I Polycarboxylate Superplasticizer (PCE) produced by Jiangsu Sobute New Materials Co.

2.2 Experimental Apparatus

The primary equipment utilized in the experiment included the ZKY-400 accelerated curing chamber, the TYA-2000 electro-hydraulic compression testing machine, the JW180 cement mortar mixer, the HJW-60 concrete mixer, and the PZT-7 vibration table. Each apparatus was employed to ensure the accurate and consistent execution of the experimental procedures, thereby enhancing the reliability and reproducibility of the results.

2.3 Mix Design

The experiment primarily investigates curing cement mortar specimens under different constant temperature conditions to observe the development of their compressive strength. Based on these data, we calculate the reaction rate constant and apparent activation energy to establish the maturity F-P equation for concrete in both positive and low positive temperature ranges.

This study focuses on developing a maturity equation applicable to concrete engineering under normal temperature and slightly above freezing conditions. Following the specification for winter construction procedures for building engineering (JGJ104-2010), this experiment sets a minimum concrete cement content of 280 kg/m³ and ensures a water–binder ratio not exceeding 0.55. What's more, the impact of fly ash addition on the reaction rate constant and apparent activation energy is also thoroughly examined. Detailed experimental results are presented in Table 2.3. These experimental set-ups and studies aim to accurately reflect the actual engineering environment to provide practically valuable research outcomes.

2.3 Mix Design

Table 2.3 Concrete mix proportion

ID	Cement (kg/m^3)	Fly ash (kg/m^3)	Water (kg/m^3)	Sand (kg/m^3)	Crushed stone (kg/m^3)
C30	380	–	185	648	1198
C40	460	–	185	590	1210
C50	510	–	178	545	1220
C30-FA	304	76	185	648	1198
C40-FA	368	92	185	590	1210
C50-FA	408	102	178	545	1220

Following the cement mortar test method recommended by ASTM C1074-11, and under the condition of maintaining consistent materials and water–binder ratios, the calculated apparent activation energy will be similar if the cement-to-sand ratio in the cement mortar is the same as the cement-to-gravel ratio in concrete. Based on this principle, we use the apparent activation energy data obtained from the cement mortar test to predict the strength development of the corresponding concrete. To make this prediction, the mix proportion of the mortar is recalculated after removing the coarse aggregate from the concrete, and the detailed data is listed in Table 2.4.

C30-FA refers to concrete where 20% of the cement mass in C30 concrete is replaced by fly ash, and similarly, C40-FA and C50-FA also utilize 20% fly ash as a replacement. M30-FA indicates concrete where 20% of the cement mass in M30 concrete is replaced by fly ash, with M40-FA and M50-FA following the same pattern.

This experiment aims to determine the compressive strength of cement mortar specimens under different curing temperatures and ages. The data obtained are analysed using a theoretical Eq. (2–1) for nonlinear fitting to derive corresponding parameters. During the initial hydration stage, the apparent activation energy of cement exhibits minimal change. When the hydration degree reaches approximately 40% to 70%, it can be considered as a constant value [1]. This is because, within this specific hydration degree range, the main stages of the cement hydration reaction are relatively stable, so the apparent activation energy does not change significantly. Based on this, and considering the prediction range of 20% to 60% of the standard compressive

Table 2.4 Cement mortar mix proportion

ID	Cement (kg/m^3)	Fly ash (kg/m^3)	Water (kg/m^3)	Crushed stone (kg/m^3)
M30	561.16	–	273.42	1769.41
M40	645.79	–	260.40	1698.60
M50	696.05	–	242.96	1665.00
M30-FA	448.93	112.23	273.41	1769.42
M40-FA	516.16	129.16	260.40	1698.59
M50-FA	556.84	139.21	242.95	1655.02

Table 2.5 Curing temperature and test age of cement mortar specimens

Curing temperature (°C)	Test age (days)
35	0.5, 1, 1.5, 2, 3
20	1, 2, 3, 7, 28
10	1, 3, 7, 21, 35

Table 2.6 Curing temperature and test age of concrete specimens

Curing temperature (°C)	Test age (days)
35	0.5, 1, 1.5, 2, 3
20	1, 2, 3, 7, 28
10	1, 2, 3, 5, 7

strength (i.e., the strength achieved after 28 days of curing under standard conditions), we determine the curing temperatures for the cement mortar specimens and the corresponding test ages under different curing conditions. Table 2.5 provides details on the specific curing temperatures and test ages under each temperature.

$$\frac{S}{Su} = \frac{\sqrt{k(t-t_0)}}{1+\sqrt{k(t-t_0)}} \qquad (2.1)$$

Considering that all necessary parameters for the concrete maturity equation can be derived using the cement mortar test method, we adopt the concrete test age setup mentioned in reference [2], with specific data provided in Table 2.6. This approach facilitates a more accurate evaluation of concrete performance and strength development.

2.4 Experimental Content and Methodology

2.4.1 *Experimental Preparation*

(1) First, thoroughly clean the concrete forming moulds ($100 \times 100 \times 100 mm$) and cement mortar specimen moulds to ensure no residues remain on their surfaces. Next, evenly apply a release agent inside both moulds to facilitate smooth demoulding. Finally, place a piece of paper at the bottom of the mould to prevent materials from leaking out during moulding, ensuring the accuracy and reliability of the test results.

(2) Before starting the experiment, conduct a comprehensive inspection and calibration of all laboratory instruments to ensure proper functioning. This step aims to prevent potential equipment failures during the experiment, avoiding material waste and errors in experimental data.

2.4.2 Experimental Methods and Procedures

(1) Preparation of cement mortar specimens.

The purpose of moulding cement mortar specimens is to determine the apparent activation energy Ea and the reference temperature T0 at two different temperature ranges: 20–35 °C and 10–20 °C. Therefore, after the completion of the cement mortar specimens, they are placed in temperature environments of 35 °C, 20 °C, and 10 °C for curing.

The specific experimental method is as follows: After selecting the materials according to the mix proportion of the cement mortar specimens listed in Table 2.4, ensure the weighing error for medium sand is within ± 1% and for cement, water, and admixtures within ± 0.5%. Then, the weighed materials are poured into a cement mortar mixer and mixed for a certain period of time. After the mixture is evenly blended, it is added to the moulds in 2–3 times. The surface is levelled using vibration moulding, ensuring that the specimen's surface aligns with the top edge of the mould. Finally, a layer of plastic wrap is placed over the surface to prevent moisture loss, and the specimens are placed in the corresponding temperature range for curing.

(2) Preparation of concrete specimens.

To determine the compressive strength of concrete, specimens are cast and cured at temperatures of 35 °C, 20 °C and 10 °C. After curing, the specimens are tested for their mechanical properties at specific ages, following relevant standards, to obtain the final data.

(3) Flexural and compressive strength test of cement mortar specimens.

For cement mortar specimens cured under different temperature conditions, samples are taken at specific ages and tested for mechanical properties following the standard procedure of the test method of cement mortar strength (ISO method) (GB/T 17,671–2021).

(4) Compressive strength test of cubical concrete specimens.

In this study, concrete specimens with dimensions of $100 \times 100 \times 100 mm$ are used. To ensure the accuracy and reliability of the test results, the mechanical properties of the specimens are tested in strict accordance with the Standard for test methods of concrete physical and mechanical properties (GB/T 50,081–2019).

2.5 Summary of the Chapter

This chapter provides a detailed description of the various raw materials used in the experiment and their basic properties, which are crucial for ensuring the accuracy and repeatability of the experiment. Through meticulous research and calculation, we determine the optimal mix proportion for cement mortar and concrete, which is key to achieving the desired performance of the specimens. Additionally, this chapter specifies the curing regime for the specimens and the specific ages for data collection, providing clear time nodes for subsequent data analysis and strength assessment. Regarding experimental equipment, we employ industry-standard devices and provide a detailed introduction to ensure the standardization of the experimental process and the reliability of the results.

References

1. Yan PY, Zheng F (2006) Kinetics model for hydration mechanism of cementitious materials. J Silic 34(5):555–559
2. Xiang ZX (1990) Study on maturity. Concrete 6:14–25

Open Access This chapter is licensed under the terms of the Creative Commons Attribution-NonCommercial-NoDerivatives 4.0 International License (http://creativecommons.org/licenses/by-nc-nd/4.0/), which permits any noncommercial use, sharing, distribution and reproduction in any medium or format, as long as you give appropriate credit to the original author(s) and the source, provide a link to the Creative Commons license and indicate if you modified the licensed material. You do not have permission under this license to share adapted material derived from this chapter or parts of it.

The images or other third party material in this chapter are included in the chapter's Creative Commons license, unless indicated otherwise in a credit line to the material. If material is not included in the chapter's Creative Commons license and your intended use is not permitted by statutory regulation or exceeds the permitted use, you will need to obtain permission directly from the copyright holder.

Chapter 3
Establishment of Maturity Equations for Different Temperature Intervals

This section focuses on studying the correlation between reaction rate constants and temperature, and examines the effects of temperature, water–binder ratio, and fly ash on the reaction rate and apparent activation energy. Additionally, maturity equations for two different temperature ranges were developed to explore the relationship between strength and maturity, thereby providing theoretical support for predicting the strength performance of concrete.

3.1 Quantitative Study of the Effect of Temperature Change on the Rate Constants of Cement Hydration Reaction

The establishment of the maturity equation is closely linked to studying the relationship between reaction rate constants and temperature. By conducting mechanical performance tests on cement mortar specimens according to the set experimental age, strength data is obtained after curing under different temperature conditions. By fitting the given age with the strength data obtained from the tests, the reaction rate constants of the cement mortar specimens under different curing conditions are determined. Then, based on the correlation between temperature and reaction rate constants recorded in the literature, their relationship is analysed to obtain the apparent activation energy. The maturity equation is established and incorporated into the prediction program as a predictive model, which has been adopted by ASTM and included in its standards.

Before determining the hydration reaction rate constants, it is necessary to establish the relationship equation between strength and time. The development of cement strength is closely related to the cement hydration process, which is divided into chemical and physical reactions. Generally, there are two different explanatory methods: the dissolution-precipitation theory defines that the hydration products formed by the slow reaction of cement dissolved in water precipitate out due to insufficient solubility, while another theory states that the mineral components of cement are insoluble in water and react directly with water to form hydration products, known as the solid-phase reaction theory [1].

Under constant temperature conditions, the strength of cement develops slowly, then quickly, and then slowly again, with an induction period in between. The end of this period signifies the final setting of the concrete, before which the cement is considered to have no strength [2]. Figure 3.1 shows the development of concrete strength. Figure 3.2 uses a hyperbolic equation to represent the simplified relationship between strength and age.

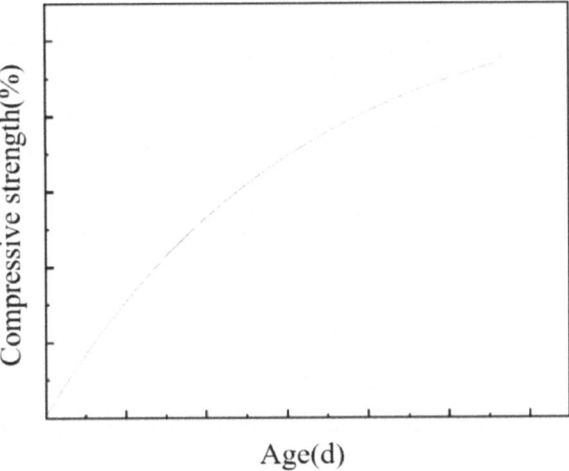

Fig. 3.1 Age of conservation–compressive strength curve [3]

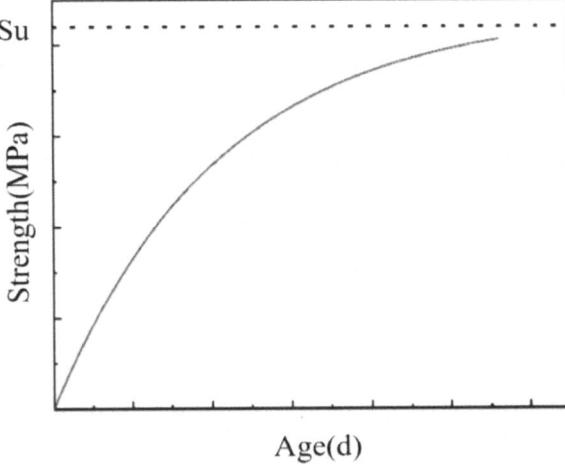

Fig. 3.2 Hyperbolic representation of strength-age relationship sketch [4]

3.1 Quantitative Study of the Effect of Temperature Change on the Rate ...

By analysing the data from the two graphs mentioned, it can be observed that the strength of concrete shows a specific trend over time. This trend bears a significant resemblance to the variation described by the hyperbolic equation proposed by Knudsen and Carino. Therefore, we have reason to infer that the hyperbolic equation can be used as an approximate model to describe the relationship between concrete strength and time. In subsequent research, we will further explore the application of the hyperbolic equation in predicting concrete strength and verify its accuracy and applicability.

$$\frac{S}{S_u} = \frac{\sqrt{k(t - t_0)}}{1 + \sqrt{k(t - t_0)}} \tag{3.1}$$

Kundsen[5] proposed eq. (3.1), which is based on the nonlinear characteristics of the cement hydration process, that is, the relationship between the degree of hydration, time, and hydration rate. Given the excellent fitting ability demonstrated by this equation, this study adopts it to explore the correlation between the compressive strength of mortar specimens and the experimental age.

Using MATLAB software, strength data obtained from mortar specimens with six different mix proportions at various experimental ages were analysed, and nonlinear regression fitting was performed using eq. (3.1). Since the curing temperature was constant, there was no need to consider relative errors in the equation regarding temperature, and the smooth curve in Figure 3.3 represents the nonlinear fitting curve.

Figure 3.3 shows a rapid increase in strength of cement mortar specimens with six mix proportions after final setting, indicating an exceptionally intense hydration reaction at this stage, which significantly enhances the strength. Through longitudinal comparative analysis, it can be observed that under the condition of increased curing temperature, the hydration reaction of cement mortar specimens with different mix proportions is more intense, leading to a sharp rise in their strength growth curves, which eventually stabilize gradually.

Using MATLAB software for nonlinear fitting of the relationship between the strength and age of cement mortar specimens, the setting time t0, reaction rate constant k, and the final strength value S_u of the cement were obtained for different mix proportions under different curing temperatures, and the results are listed in Table 3.1.

Analysis of the reaction rate constants for the six mix proportions of cement mortar under different curing temperatures in Table 3.1 reveals the following:

For cement mortar specimens with the same components but different water–binder ratios, the reaction rate constant increases as the water–binder ratio decreases under the same curing temperature. For specimens with the same mix proportion, their reaction rate constants increase with the rise in curing temperature. The reaction rate constants of cement mortar specimens containing fly ash are slightly lower than those without fly ash.

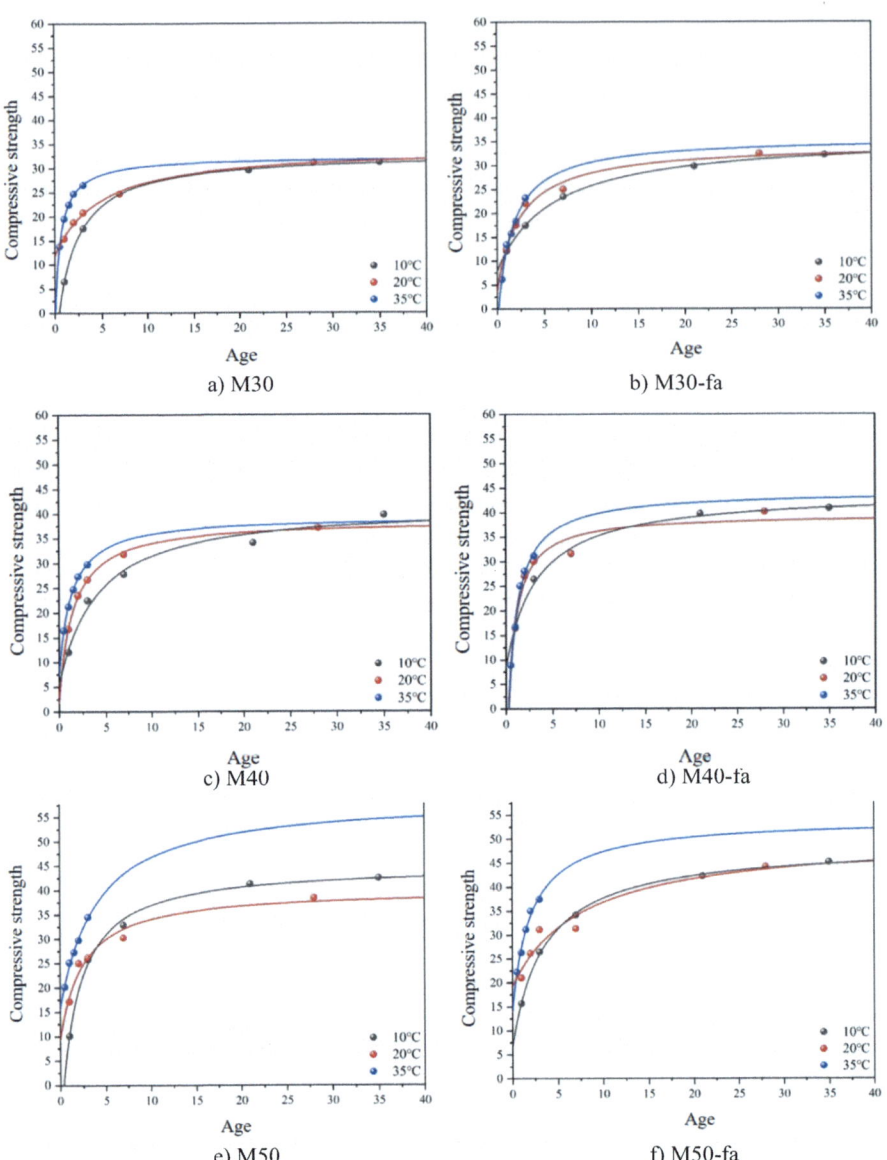

Fig. 3.3 Fitted relationship between compressive strength of cement mortar specimens of different mixing proportions with age change

3.1 Quantitative Study of the Effect of Temperature Change on the Rate …

Table 3.1 t_0, k, S_u for the required mortar specimens

Type	T	S_u	Standard deviation	k (1/d)	Standard deviation	t_0	Standard deviation	R^2
M30	35°C	32.36184	7.00234	0.75436	0.20138	0.23403	0.04588	0.9989
	20°C	34.26114	2.99886	0.32683	0.06183	0.28726	0.47024	0.99625
	10°C	33.04531	0.42389	0.21301	0.01759	0.85777	0.03191	0.99978
M30-FA	35°C	32.71522	8.57424	0.57074	0.10753	0.39968	1.35256	0.9947
	20°C	34.35557	2.11263	0.37108	0.16984	0.45581	0.50037	0.9814
	10°C	36.03593	0.33954	0.12334	0.01188	1.25968	0.12639	0.99972
M40	35°C	39.45248	3.89249	0.97996	0.19464	0.22758	0.09998	0.99805
	20°C	38.64191	5.58745	0.33341	0.07755	0.36458	0.1211	0.99821
	10°C	41.76162	1.0718	0.22962	0.12357	0.60909	0.73888	0.97879
M40-FA	35°C	43.67849	6.65208	0.83547	0.31338	0.23032	1.5489	0.96499
	20°C	39.60112	3.08734	0.30758	0.63529	0.31691	0.42794	0.94585
	10°C	44.04842	1.42674	0.17867	0.08451	0.38096	0.39073	0.99392
	35°C	55.02081	9.11531	0.94505	0.24794	0.14296	0.55821	0.99087
M50	20°C	40.19456	3.09407	0.32645	0.27644	0.36586	0.84753	0.9631
	10°C	45.0132	1.17942	0.27339	0.07493	0.87531	0.13696	0.99644
	35°C	54.12206	8.36987	0.97248	0.47043	0.11352	0.43902	0.98319
M50-FA	20°C	50.44945	1.56397	0.31022	0.20996	0.21625	3.4099	0.92888
	10°C	48.67457	1.74828	0.18815	0.03113	0.47211	0.15859	0.9989

To visually display the trend of reaction rate constants with temperature changes and the impact of different mix proportions and additives, Figure 3.4 presents a scatter plot with temperature (T) on the horizontal axis and reaction rate constant (k) on the vertical axis, covering six different mix proportions. This graph aims to precisely reveal the extent of temperature's impact on the reaction rate constants and the characteristics of changes under different conditions.

In the temperature range of 10–35 °C, the reaction rate constants of cement mortar specimens with the same water–binder ratio but different components are similar. The reaction rate constants of specimens containing fly ash are slightly lower than those without fly ash because, in the early stages of hydration, fly ash hardly participates in the reaction, resulting in slightly lower constants than the control group [6].

In the field of predicting concrete strength maturity equations, exploring the impact of temperature changes on reaction rate constants is particularly crucial. The research by Freiesleben and Pedersen [7] indicated that the sensitivity of the reaction rate to temperature increases with rising temperature and this dependency exhibits an exponential change characteristic. Based on this insight, this study adopted MATLAB software to perform linear and exponential fitting analyses on the relationship between reaction rate constants and temperature, as shown in Figure 3.5,

Fig. 3.4 Scatter plot of reaction rate constant k versus temperature T

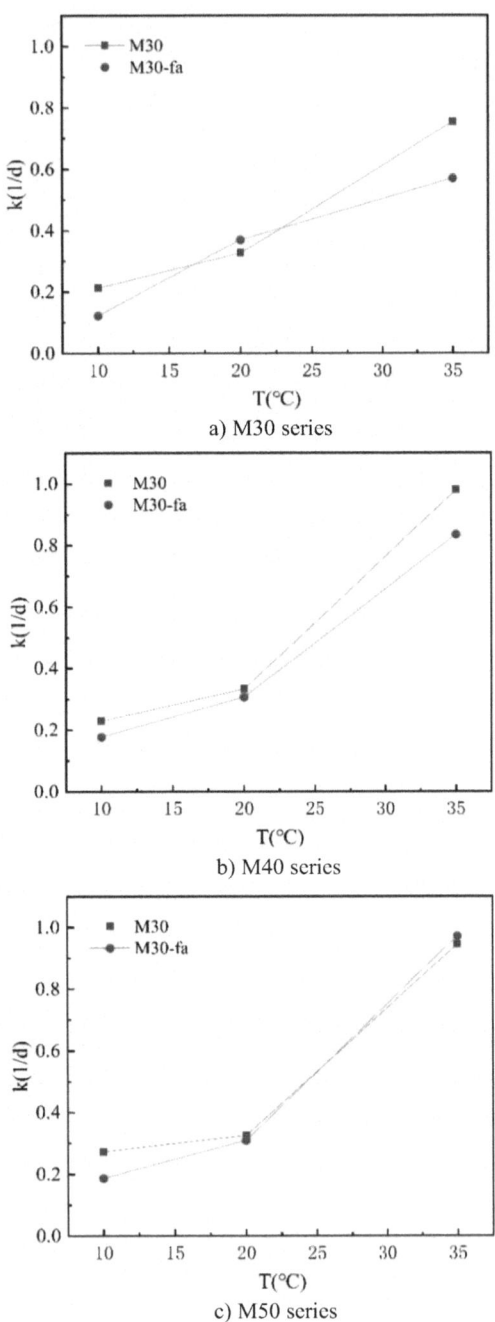

a) M30 series

b) M40 series

c) M50 series

3.1 Quantitative Study of the Effect of Temperature Change on the Rate … 29

aiming to deepen the understanding of the patterns of reaction rate constants with temperature changes.

The graph shows that for cement mortar specimens with different mix proportions, the fitting of data points using an exponential equation is superior to linear fitting. This indicates that the hydration reaction rate is highly sensitive to temperature changes;

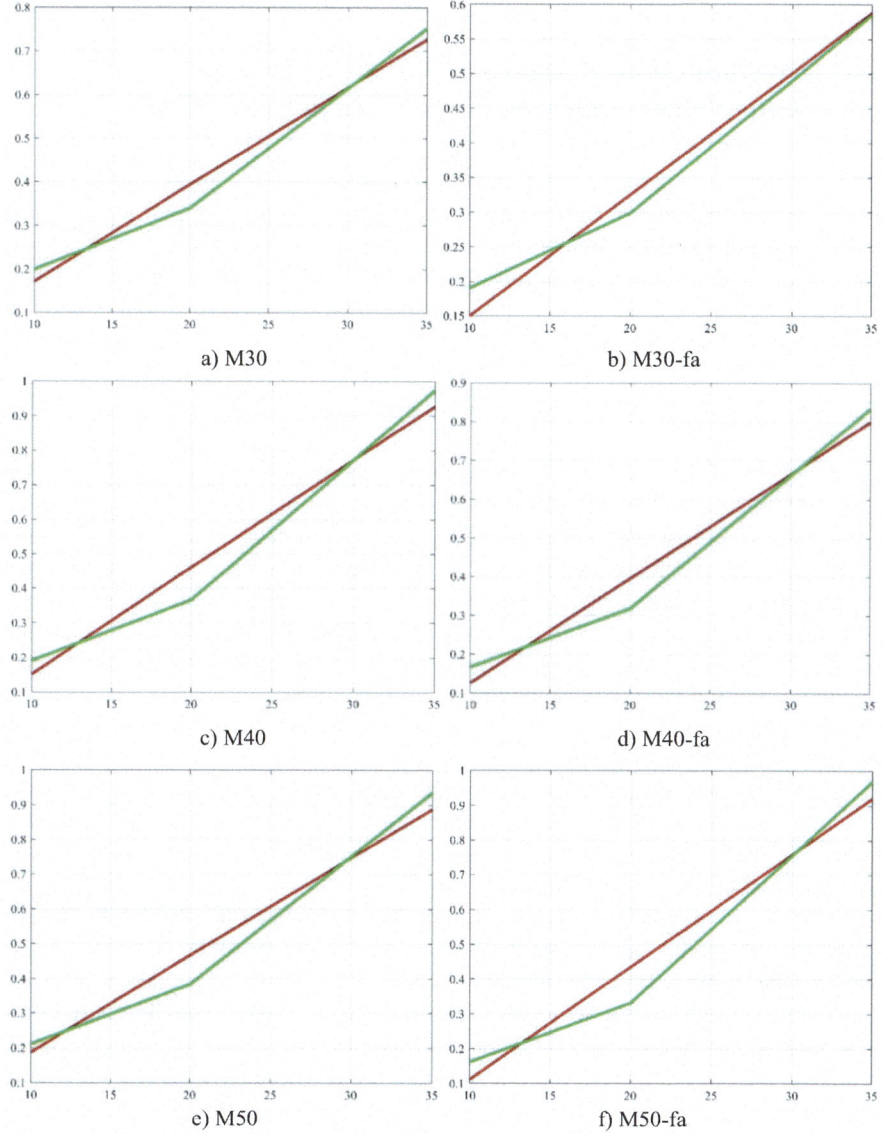

Fig. 3.5 k versus T fit plot

the impact on the reaction rate constant is greater at higher curing temperatures, while changes in the reaction rate constant are minimal at lower temperatures.

3.2 F-P Maturity Model Construction Based on Specific Temperature Intervals

3.2.1 Determination of Apparent Activation Energy in the Positive and Low Positive Temperature Range

Based on extensive experimental data analysis, Arrhenius proposed a theory suggesting that the relationship between chemical reaction rates and temperature is not linear, but exhibits geometric multiple change. This led him to develop an expression that describes the relationship between temperature and reaction rates, known as eq. (3.2). This discovery provides an important theoretical foundation for understanding chemical kinetics [8].

$$k(T) = A \cdot e^{-\frac{E_a}{RT}} \tag{3.2}$$

The concept of activation energy explains this equation well. The prerequisite for a chemical reaction is the collision between molecules, and the excess energy after the collision in an activated reaction is called activation energy [9]. Understanding this energy is crucial for predicting the early strength development and process of concrete. Over the years, with advancements in science and technology, the concept of activation energy has been further explored and gradually introduced into cement chemical reactions. However, to differentiate it from the traditional chemical reactions, the activation energy in cement hydration reactions is referred to as apparent activation energy, which is used to represent the impact of different temperature conditions on the hydration reaction rate.

Carino [10] proposed a method to accurately determine the apparent activation energy of concrete with specific components. This method has been officially adopted by ASTM (American Society for Testing and Materials) and incorporated into relevant standards. The steps are as follows:

First, the corresponding groups of mortar specimens (40×40×160mm) are prepared at room temperature, with each group containing three specimens to ensure representative results. Then, to simulate the curing conditions in actual engineering projects, the cement mortar specimens are sealed with plastic wrap and cured in a constant temperature environment. When the specimens reached the predetermined curing age, they are removed from the constant temperature environment, and their compressive strength is immediately tested. The obtained strength and time data are fitted and processed using MATLAB software. This process successfully establishes the relationship between compressive strength and the reaction rate constant, allowing for the calculation of the reaction rate constant value. Finally, based on the

3.2 F-P Maturity Model Construction Based on Specific Temperature Intervals

principle of the Arrhenius equation, the relationship between the logarithm of the reaction rate (lnk) and the reciprocal of the curing temperature (expressed in Kelvin, 1/T) is further fitted. Partial MATLAB code for the fitting process is provided in Appendix B.

Taking logarithms on both sides of eq. (3.2) simultaneously yields the following eq. (3.3):

$$\ln k(T) = \ln A - \frac{E_a}{RT} \qquad (3.3)$$

Apparent activation energy reflects both the sensitivity of the reaction rate to temperature and the difficulty of the activation process. A higher activation energy indicates a more difficult reaction.

By taking the logarithm of the reaction rate constants and the reciprocal of the corresponding temperatures from Table 3.1, and applying linear regression to the above equation, we can obtain the constants A and Ea/R. Using MATLAB software to process the data of cement mortar with six mix proportions, the apparent activation energy (Ea) can be calculated. The results are listed in Table 3.2.

The apparent activation energy obtained from the two temperature ranges shows that it decreases with increasing temperature. The water–binder ratio has no significant effect on the apparent activation energy. The apparent activation energy of cement mortar specimens incorporating fly ash is slightly higher than that of specimens without fly ash. This can be explained from the kinetics and physical chemistry of cement hydration. As cement hydration is a complex chemical reaction process, its reaction rate has more dependence on the temperature, and according to the Arrhenius equation mentioned above, the reaction rate increases with increasing temperature. Higher temperatures provide more energy for the reaction, enabling more reactant molecules to reach or exceed the activation energy. Apparent activation energy is essentially a parameter reflecting energy barriers, and at higher temperatures, it requires less energy to complete the reaction. From a physicochemical change perspective, as the temperature rises, some compounds in the cement, such as silicates and aluminates, react faster, usually with lower activation energies.

Table 3.2 Correlation data obtained by linear fitting of lnk to 1/T

Cement mortar	10–20 °C		20–35 °C	
	lnA	E_a (kJ/mol)	lnA	E_a (kJ/mol)
M30	20.19	50.24	19.23	36.27
M30-fa	24.56	57.20	11.29	43.38
M40	18.29	58.30	15.30	43.20
M40-fa	25.10	55.96	14.98	39.89
M50	18.93	43.98	18.20	49.22
M50-fa	19.03	46.90	17.85	44.99

Additionally, the increase in temperature may alter the microstructure of the cement paste, thereby reducing the energy barriers needed for hydration reactions [11–13].

In this chapter, the author analyses factors affecting apparent activation energy, with the main purpose of determining the equivalent age of concrete after obtaining the apparent activation energy to predict the final compressive strength.

3.2.2 Relationship Between Equivalent Age and Compressive Strength

To establish the F-P maturity equation, it is necessary to investigate the relationship between compressive strength and equivalent age. This allows for the prediction of strength development using the established maturity equation. Based on the reaction rate constants at different temperatures obtained in Sect. 3.1 and the curing conditions of the cement mortar specimens[14], the actual ages at these temperatures are converted into equivalent ages (t_e) at a reference temperature of 20 °C using eq. (3.4). Equation (3.5) is derived by substituting the k(T) function from eq. (3.4). Once the equivalent age t_e is calculated, the strength equation at this equivalent age (3.7) is obtained by comparing the development eq. (3.6) of the temperature 20°C with time to obtain the calculated strength of the concrete.

$$t_e = \frac{k(T)}{k_r} \cdot \Delta t \tag{3.4}$$

$$t_e = e^{-E_a\left(\frac{1}{T} - \frac{1}{T_r}\right)} \cdot \Delta t \tag{3.5}$$

$$\frac{S}{S_u} = \frac{\sqrt{k(t - t_0)}}{1 + \sqrt{k(t - t_0)}} \tag{3.6}$$

$$S_{js} = S_u \frac{\sqrt{k(t_e - t_0)}}{1 + \sqrt{k(t_e - t_0)}} \tag{3.7}$$

To ensure that the actual strength development of the cement mortar specimens at different equivalent ages aligns with the strength-time relationship represented by Equation (3.6), a scatter plot (Figure 3.6) is generated, showing the relationship between the measured compressive strength and equivalent age. This confirms that the calculated strength aligns with the strength predicted by the F-P maturity method. The smooth curve represents the strength development curve at 20 °C.

As shown in the figure, the measured strength values at different temperatures align with the strength development trend at the reference temperature of 20 °C for the same equivalent age. This confirms that the actual strength development of the

3.2 F-P Maturity Model Construction Based on Specific Temperature Intervals

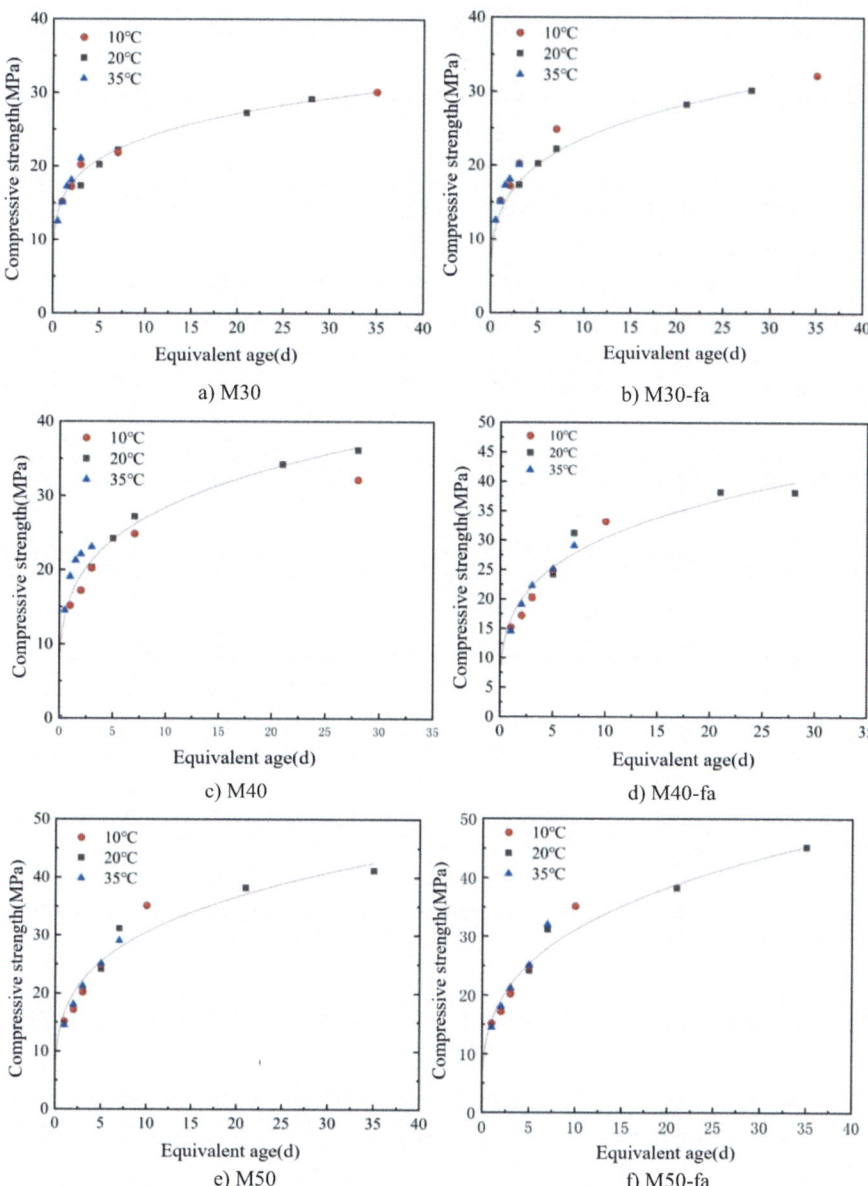

Fig. 3.6 Scatter plot of measured compressive strength of cement mortar versus equivalent age

mortar specimens with increasing equivalent age follows the strength-time relationship represented by Equation (3.7), which is consistent with the relevant provisions in ASTM C1074-11. Therefore, the predicted strength values can be calculated.

3.3 Analysis of the Prediction Accuracy of the F-P Maturity Equation

The F-P maturity equation established within the temperature range of 10–35 °C requires an assessment of its prediction accuracy. Using the apparent activation energy and equivalent age calculated in sects. 3.2.1 and 3.2.2, the predicted strength is calculated through the conversion relationship between equivalent age and strength. This is compared with the actual strength obtained from specimens of six different mix proportions. The method of predicting the corresponding concrete strength development using the apparent activation energy of mortar specimens has been adopted by ASTM and included in the standard [15].

The method of using the apparent activation energy of the mortar specimens to predict the strength development of the corresponding concrete is as follows:

(1) Fit the concrete strength data under curing at the reference temperature with the experimental age using MATLAB software, adopting the strength-age relationship eq. (3.6) to obtain the reaction rate constant k(T) under reference curing temperature;
(2) Insert the apparent activation energy obtained in sect. 3.2.1 into eq. (3.8) and calculate the equivalent age t_e at the reference temperature for multiple curing temperatures using MATLAB software.
(3) Insert the converted equivalent age t_e into the relationship eq. (3.7) in Sect. 3.2.2 to obtain the calculated strength values for each experimental age at multiple curing temperatures.

$$t_e = \int_{t_0}^{t} e^{\frac{-E_a}{R}(\frac{1}{T} - \frac{1}{T_r})} dt \quad (3.8)$$

Since the reference temperature used in this experiment is 20 °C, nonlinear fitting is performed for the strength–age relationship at 20 °C for C30, C30-fa, C40, C40-fa, C50, C50-fa, as shown in Table 3.3.

The F-P maturity equation was selected based on the selected temperature interval:

Table 3.3 Su, k, t0 of C40, C40-fa concrete cured at constant temperature at 20 °C

Concrete	Ultimate strength S_u(MPa)	Reaction rate constant k (1/d)	Final setting time t_0(d)	Correlation coefficient R^2
C40	46.34	0.350	0.298	0.925
C40-fa	49.30	3.352	0.321	0.916

$$t_e = e^{-E_a(\frac{1}{T} - \frac{1}{T_r})} \cdot \Delta t \tag{3.9}$$

To visually assess the prediction performance of the theoretical model, a scatter plot 3–7 is drawn with the actual strength of the specimens on the horizontal axis and the strength calculated by the theoretical equation on the vertical axis (Fig. 3.7).

From the analysis of the above graph, the error rate between the actual strength of most specimens and the strength predicted by the model is controlled within ±10%. To directly observe the absolute error between the predicted strength and the actual strength within the temperature range, an absolute error comparison chart 3-8 of concrete under different curing temperatures is drawn (Fig. 3.8).

From the above graph, it is evident that the absolute error range for C40 concrete is between 1 and 6%, the same as for concrete with different mix proportions. Within the temperature range of 10–35 °C, the F-P maturity equation demonstrates a relatively low level of absolute error, indicating its good predictive performance. This important finding not only verifies the applicability of the F-P maturity equation under specific temperature conditions but also lays a solid theoretical foundation for the subsequent construction of intelligent prediction programs.

3.4 Improving the F-P Maturity Equation

To improve the F-P maturity equation from a methodological perspective, the following aspects can be considered:

1. Enhance data quality and quantitative indicators to increase the accuracy and quality of data collection, considering the impact of the two temperature ranges described above on the F-P maturity equation.
2. Utilize MATLAB software to adopt calculations of equivalent age and apparent activation energy from other articles, using code or integrated learning methods to enhance the stability and accuracy of the F-P maturity equation.

Since the F-P maturity equation needs to be integrated into a concrete strength intelligent prediction program, relevant theoretical calculation methods can be used. By calculating the potential energy surface of the reaction through quantum chemical methods, the activation energy can be estimated to subsequently obtain the corresponding equivalent age. This method is based on the need for substantial computational resources and requires a high understanding of the reaction mechanism.

From the calculations summarized in the above sections, it is evident that predicting concrete strength using the F-P maturity equation requires the hydration reaction rate constant k, and it is crucial to obtain the apparent activation energy E_a, which is then converted into the equivalent age t_e. To better integrate the F-P maturity equation with artificial neural network predictions of concrete strength, it is necessary to quickly obtain the equivalent age t_e. The nonlinear relationship between k and T referenced [16] ensures rapid calculation of te through MATLAB software.

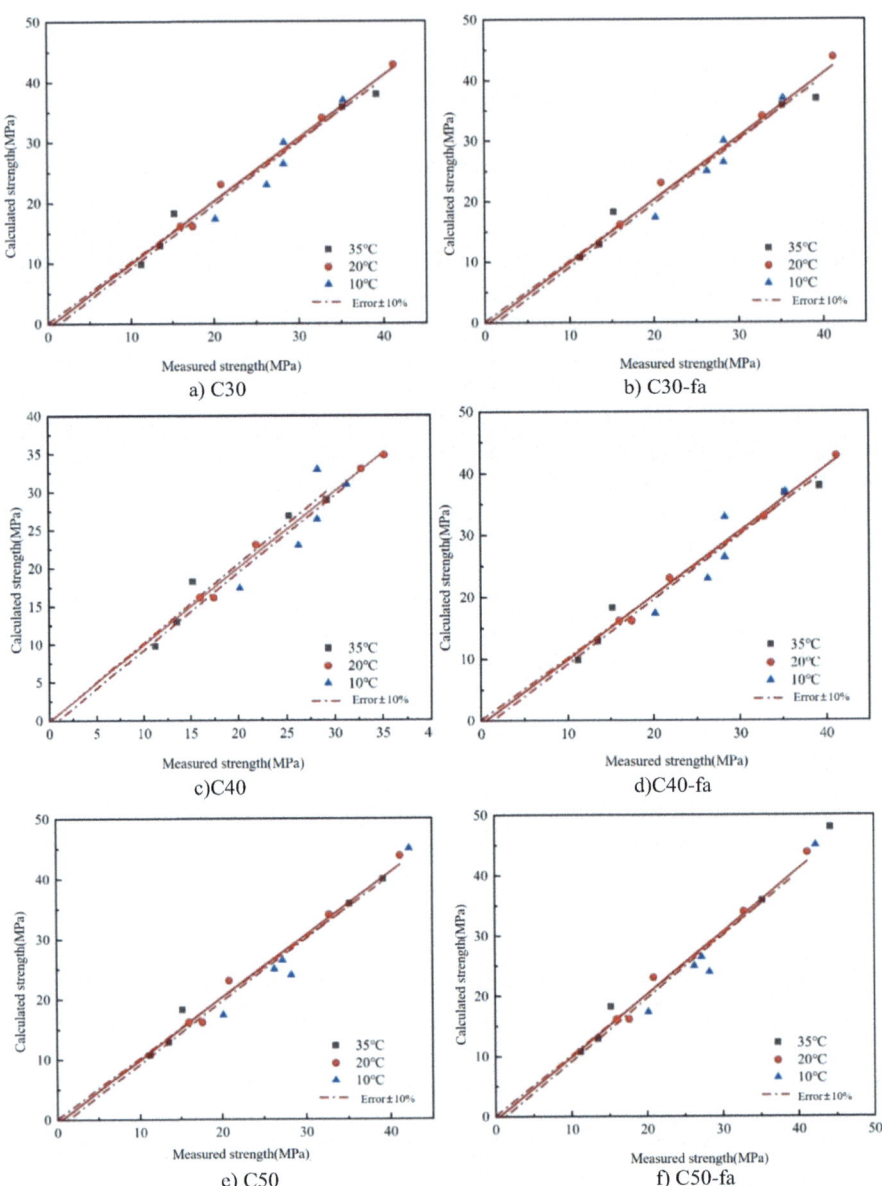

Fig. 3.7 Comparison of calculated and measured strengths

3.4 Improving the F-P Maturity Equation

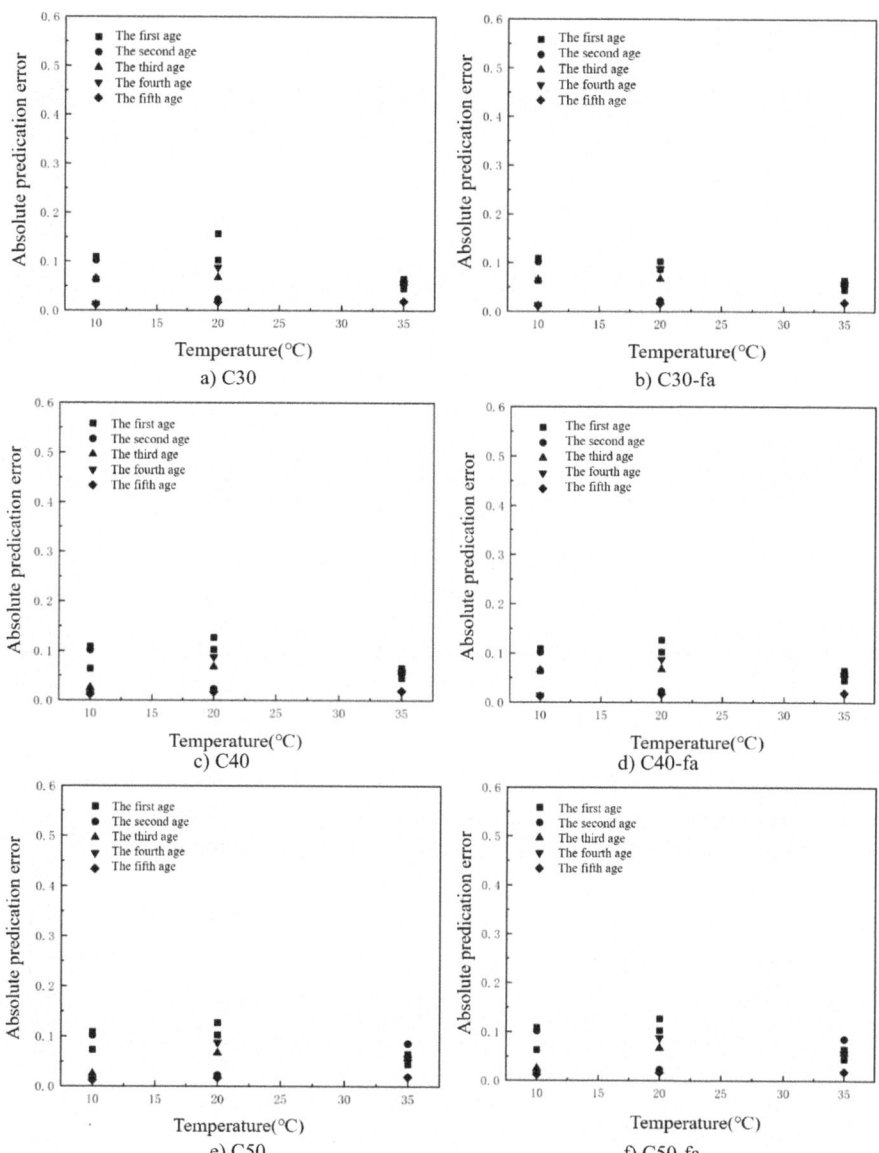

Fig. 3.8 Absolute prediction error of concrete at different curing temperatures

The following relation used:

$$t_e = 0.1132 + 0.03446T + 0.000548T^2 \tag{3.10}$$

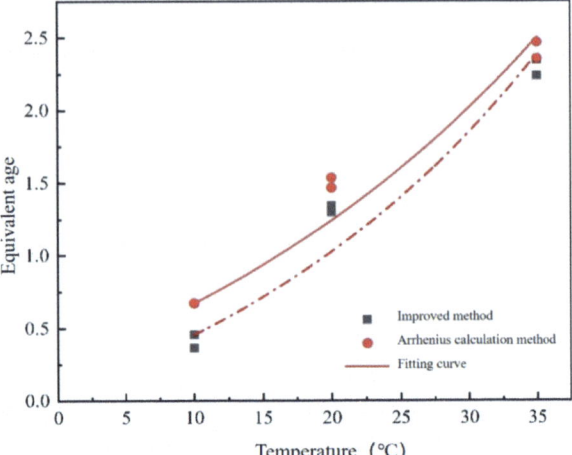

Fig. 3.9 Comparison of the curves of the improved method and the Arrhenius calculation method [17]

To visually verify the relationship between this equivalent age and temperature aligns with the described algorithms, the fitting curves of both methods are compared as shown in Figure 3.9.

Thus, it is also evident that the method of obtaining the apparent activation energy E_a through theoretical calculations, and then calculating the key parameter equivalent age t_e, is applicable and fits well with subsequent integration into intelligent prediction programs.

The equation proposed in the references has an error within ±6% between 10 and 35 °C, which adequately prepares for the integration of the F-P maturity equation into intelligent prediction programs for predicting concrete strength.

3.5 Summary of the Chapter

This chapter primarily explores the establishment process of the F-P maturity equation, clarifies the relationship between the reaction rate constant and temperature changes, and studies the effects of temperature, fly ash, and water–binder ratio on the subsequently obtained reaction rate constants and apparent activation energy. The MATLAB software was used to fit the data and quickly obtain the required equivalent age to improve the F-P maturity equation. The conclusions obtained are as follows:

(1) The reaction rate constant changes nonlinearly with temperature. For cement mortars with the same water–binder ratio, those containing fly ash have a lower reaction rate constant than those without fly ash. For mortars with different water–binder ratios, under the same environmental conditions, the reaction rate constant decreases as the water–binder ratio increases.

(2) In the temperature ranges of 10–20 °C and 20–35 °C, there is no significant correlation between the apparent activation energy and the water-to-cement ratio, but an increase in temperature reduces the apparent activation energy. With the same water–binder ratio, the apparent activation energy of cement mortars containing fly ash is higher than that of those without fly ash.

(3) In this chapter, we successfully established a set of equations based on the F-P maturity theory aimed at accurately predicting the development of concrete compressive strength. This is significant for the subsequent development of intelligent prediction programs and becomes an indispensable part of the core calculation process. Although the maturity model we established meets the precision requirements of engineering applications, its calculation process is complex, from determining the apparent activation energy E_a to converting it to equivalent age t_e, each step requires meticulous data processing and analysis. In view of this, we propose to quickly obtain the equivalent age t_e in the F-P maturity equation using an optimized relationship to accelerate concrete strength prediction, with the error control of the optimized maturity equation within ±6%.

References

1. Shen X, Guo SH, LI WW, et al (2023) Research status on hydration and properties of low-heat Portland cement. Bull Chinese Ceram Soc 42(2)
2. Zhang WH, Zhang YS (2015) Research progress on the hydration, hardening and microstructure formation mechanism of modern concrete under high temperature curing conditions. Bull Chinese Ceram Soc 34(01):149–155
3. Wang JC, Yan PY (2014) Influence of temperature history on compressive strength of early age concrete. J Northwest A&F Univ 42(7):228–234
4. Zhong YH, Wu L, Liu CF et al (2021) Early age strength test and maturity analysis of low heat cement concrete. Yangtze River 52(09):186–192
5. Fava G, Ruello ML, Corinaldesi V (2011) Paper mill sludge ash as supplementary cementitious material. J Mater Civil Eng 23(6):772–776
6. Zhang YS, Sun W, Zheng KR et al (2006) Hydration process of Portland cement-fly ash pastes. J Southeast Univ (Nat Sci Edn) 36(1):118–123
7. Maruyama I, Lura P (2019) Properties of early-age concrete relevant to cracking in massive concrete. Cem Concr Res 123:105770
8. Dong H, Chen X, Yang G, et al (2024) Rheological model of cement-based material slurry with different water-cement ratio and temperature. Multidiscipline modeling in materials and structures. J Rheol Model 20(1):159–177
9. Li BL, Ren WR, Cui CF et al (2023) Application of closed-loop-experiment-teaching mode in education of general chemistry: measurement of chemical reaction rate constant and activation energy. Chem Educ (in Chinese and English) 44(24):82–86
10. Ferreira L, De Brito J, Saikia N (2012) Influence of curing conditions on the mechanical performance of concrete containing recycled plastic aggregate. Construct Build Mater 36:196–204
11. Liao YS, Gui Y, Shen Q et al (2018) Determination of apparent activation energy of hydration reaction of calcium sulfoaluminate cement. J Build Mater 21(6):864–870

12. Thomas JJ (2012) The instantaneous apparent activation energy of cement hydration measured using a novel calorimetry-based method. J Am Ceram Soc 95(10):3291–3296
13. Zhao KY, Zhang P, Kong XM, et al (2022) Recent progress on Portland cement hydration kinetic models and experimental methods. J Chinese Ceram Soc 50(6):1728–1761
14. Cao Y, Wang RW, Ou DF (2023) Research on estimation of concrete strength for winter construction based on maturity theory. Highway 68(09):367–373
15. Carino NJ, Lew HS (2001) The maturity method: from theory to application. In: A structural engineering odyssey, pp 1–19
16. Ma Y, Ma L, Pan BL (2019) Research on predicting early strength of concrete based on maturity method. Technol Highway Trans (Applied Technology Edition). 5:1–4
17. Zhang R, Shi N, Huang D (2013) Influence of initial curing temperature on the long-term strength of concrete. Magaz Concr Res 65(6):358–364

Open Access This chapter is licensed under the terms of the Creative Commons Attribution-NonCommercial-NoDerivatives 4.0 International License (http://creativecommons.org/licenses/by-nc-nd/4.0/), which permits any noncommercial use, sharing, distribution and reproduction in any medium or format, as long as you give appropriate credit to the original author(s) and the source, provide a link to the Creative Commons license and indicate if you modified the licensed material. You do not have permission under this license to share adapted material derived from this chapter or parts of it.

The images or other third party material in this chapter are included in the chapter's Creative Commons license, unless indicated otherwise in a credit line to the material. If material is not included in the chapter's Creative Commons license and your intended use is not permitted by statutory regulation or exceeds the permitted use, you will need to obtain permission directly from the copyright holder.

Chapter 4
Concrete Strength Prediction Based on Artificial Neural Networks

Based on artificial neural network theory, this chapter establishes a compressive strength prediction model using BP neural networks, with input parameters such as cement, water, coarse aggregate, fine aggregate, etc., and concrete strength as the output parameter. To enhance prediction precision and accuracy, optimization algorithms including Particle Swarm Optimization (PSO-BP) and Ant Colony Optimization (ACO-BP) were respectively implemented. Based on the comparative results, we ultimately integrated PSO-BP into the intelligent concrete strength prediction model.

4.1 Establishment of BP Neural Network Model

4.1.1 BP Neural Network Models

The concrete mix proportions and compressive strength data employed in this study were sourced from the database in reference [1], containing 1,030 original data samples. After excluding specimens with excessively long curing periods and those containing excessive admixtures, 1,000 valid datasets remained. This database systematically compiles relevant compressive strength data from domestic and international sources, forming a comprehensive system. Given the complex nonlinear relationships between numerous influencing factors of concrete compressive strength and mix proportion components [2], the BP neural network offers significant advantages in handling such complex nonlinear relationships.

The computational process within the BP neural network involves summing weighted input values with input layer biases, as shown in Eq. (4.1), which are then transmitted to the hidden layer.

$$net_j = \sum_{i=1}^{n} \omega_{ij} x_i + bias_j \qquad (4.1)$$

where x_i and y_i are input and output values respectively; ω_{ij} and $bias_j$ are weights and thresholds, respectively.

And where a neuron is equivalent to an independent function, its matrix form is adopted for modelling, as Eq. (4.2).

$$Y = \theta(W \cdot X + b) \qquad (4.2)$$

$$W = [W_1 W_2 \ldots W_n], X = \begin{bmatrix} X_1 \\ X_2 \\ \vdots \\ X_n \end{bmatrix}.$$

The activation function (θ), which is used as a simulation of the input or output state of a neuron, can be expressed using several different activation functions, and the following activation functions are commonly used:

Hard limiter: $\theta(a) = \begin{cases} 0, a < 0 \\ 1, a > 0 \end{cases}$

Saturated linear function: $\theta(a) = \begin{cases} 0, a < 0 \\ a, 0 \leq a \leq 1 \\ 1, a > 0 \end{cases}$

Curve function: $\theta(a) = \frac{1}{1+e^{-a}}$.

Hyperbolic tangent function: $\theta(a) = \frac{e^a - e^{-a}}{1+e^{-a}}$.

While using the common functions above, various non-computable biases should be taken into account. Based on the neural network optimized by the algorithm used in this study, and thus the tan-sigmoid function process within the neuron used is shown in Eq. (4.3) [3].

$$y_j = \theta(net_j) = (1 + e^{(-2net_j)}) - 1 \qquad (4.3)$$

4.1.2 BP Neural Network Design Process

The specific learning steps of the BP neural network are shown in Fig. 4.1.

The MATLAB's built-in app module is used to create a basic BP neural network. After loading the collected dataset and setting the corresponding parameters, the

4.1 Establishment of BP Neural Network Model

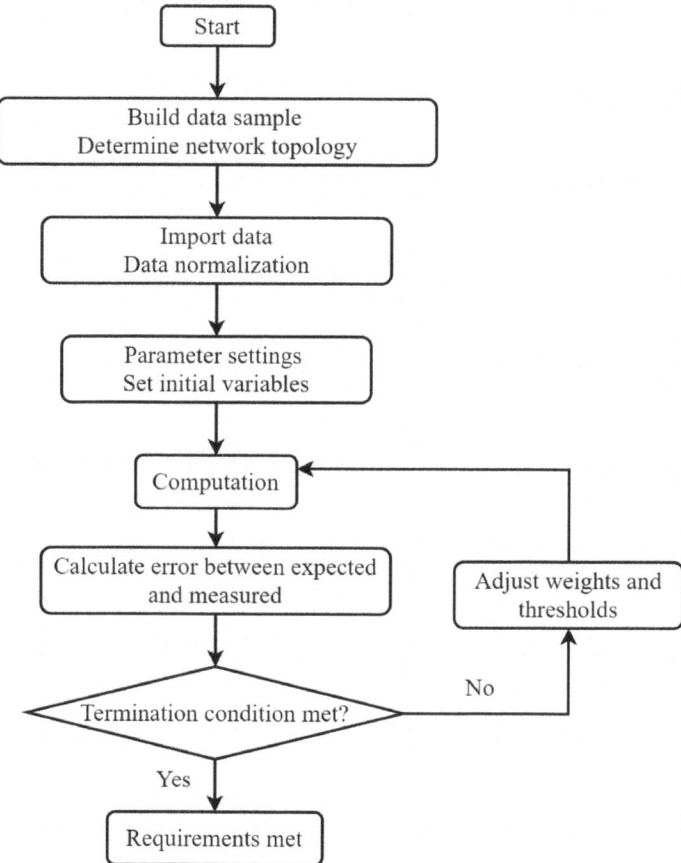

Fig. 4.1 BP neural network flowchart

software will automatically handle training, validation, prediction, and performance evaluation. The basic steps are:

(1) Import the collected dataset, determine the input and output parameters, and combine experimental data from Sect. 2.1 and relevant concrete reference data to import into the data interface;
(2) Randomly shuffle the pre-divided dataset in Sect. 4.1.1 for training, data fitting and validating the final results;
(3) Determine the number of neurons in the hidden layer;
(4) Start training, using the software's default algorithm. Once started, the training interface will display information such as iteration count and time. The Levenberg–Marquardt algorithm built into MATLAB is adopted.

When building the neural network, the selection of the number of hidden layer nodes is crucial. Too few neurons will result in poor training performance, failing to achieve the desired effect. On the other hand, too many neurons may improve accuracy to some extent but also increase computation and lead to overfitting. Therefore, it is particularly important to choose the number of hidden layer while meeting the accuracy requirements [4].

Several commonly used empirical formulas to calculate the number of hidden layer neurons are as follows: [5, 6].

$$m = \sqrt{r+l} + a \quad (4.4)$$

$$m = \log_2 r \quad (4.5)$$

$$m = \sqrt{rl} \quad (4.6)$$

In these formulas, m represents the number of hidden function neurons; r is the number of input neurons; l is the number of output neurons; and a is a constant between [1, 10].

These formulas are empirical rules, typically using a multiple of the number of input features as the number of hidden layer neurons. However, selecting just one formula can only serve as a starting point, and subsequent calculations need to be adjusted according to the specific situation. Therefore, this paper refers to the three formulas to calculate the appropriate number of hidden layer neurons.

4.1.3 BP Neural Network Input and Output Layers

When establishing a neural network, the selection of factors for the input and output layers is crucial for the performance of the predictive strength model. The output layer should be compressive strength, and when choosing variables for the input layer, some basic parameters of concrete should be considered, including the mix proportion and age. The factors chosen in this study mainly include: cement, blast furnace slag, fly ash, water, water reducer, coarse aggregate, fine aggregate, and age.

These eight parameters are chosen as influencing factors due to the prevalence of ordinary hardened concrete in engineering. Whether from the perspective of raw materials or production technology, these parameters are suitable for optimization and refinement during the self-learning process of the neural network. When using reference data from others as a database, to reduce the complexity of the model, curing conditions and service environment are not considered, and only the age issue is considered. Since the main purpose of this study is to obtain early strength, thus the strength of different ages greatly aids the model fitting process. Finally, the number of neurons in the input layer, output layer, and hidden layer are determined to be 8, 1, and 9, respectively. The dataset is divided into two parts: a training set (900 samples) and a testing set (100 samples).

4.1 Establishment of BP Neural Network Model

4.1.4 Results and Analyses

The performance of the neural network is primarily evaluated using mean squared error (MSE), where smaller values indicate higher fitting accuracy. As shown in Fig. 4.2, the MSE for both training and testing converges to around 0.02 and 0.03 by approximately the 24th epoch, demonstrating effective model training and prediction due to its simple structure. Figure 4.3 reveals correlation coefficients R^2 of 0.97118 (training) and 0.95734 (testing), with an overall dataset coefficient R^2 of 0.96664, confirming the BP neural network's successful training. The strength prediction comparison in Fig. 4.4 achieves a correlation coefficient of 0.86154, further validating its predictive capability.

However, due to the simple structure of the BP neural network, its prediction accuracy is not optimal when handling complex nonlinear problems. Additionally, it adopts a local search method that follows the gradient descent direction, which causes it to get stuck in local optima during training, learning, and prediction. This means that while the training may reach a local minimum at certain points, it does not reach the global minimum across the entire dataset, leading to variations in the model structure with each training run and preventing the optimal solution from being achieved. As a result, the prediction is more conservative and only serves as a reference.

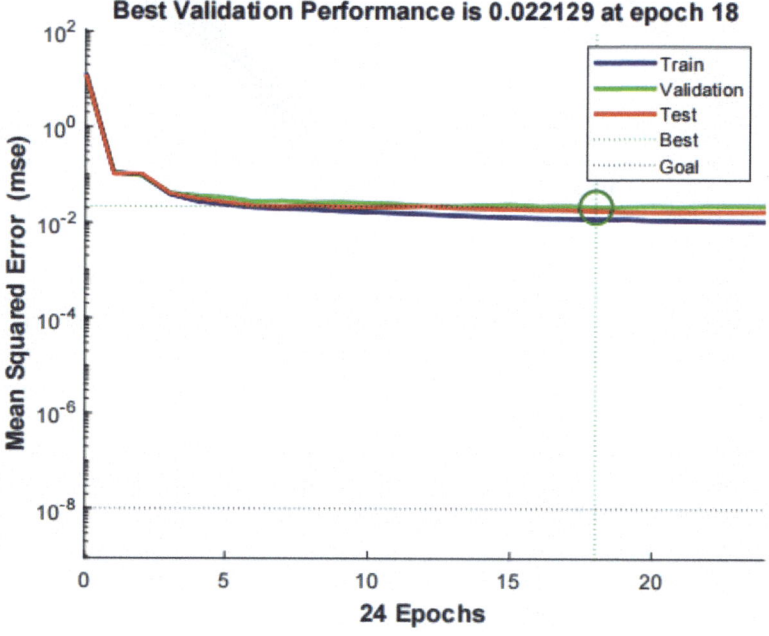

Fig. 4.2 Training and testing process mean square error

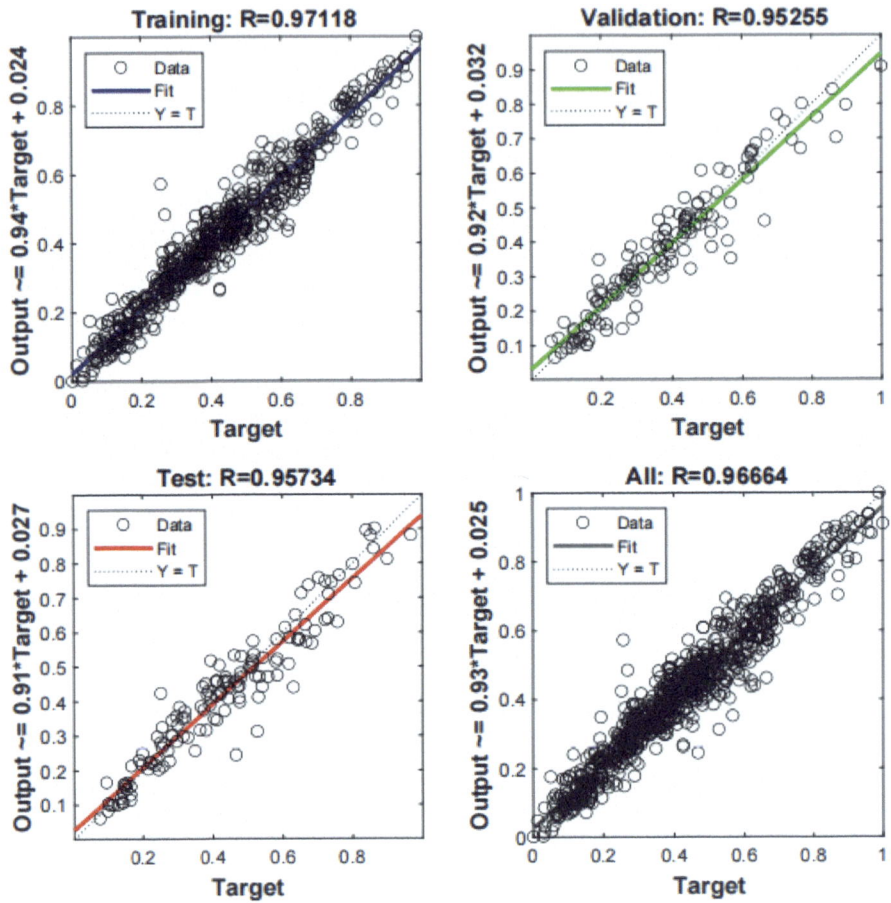

Fig. 4.3 BP neural network model training and prediction results

The basic structure of the BP neural network is mostly based on others' empirical formulas, lacking systematic and logically sound calculation steps. As a result, when overfitting occurs, the model may fail to converge, and the prediction results may not meet the expected standards.

To address the limitations of the BP neural network, this study employs two different optimization algorithms—particle swarm optimization (PSO) and ant colony optimization (ACO)—to compare their optimization efficiency. From a global perspective, these algorithms aim to compensate for the shortcomings of traditional BP neural networks and identify the most suitable optimization algorithm for predicting concrete strength.

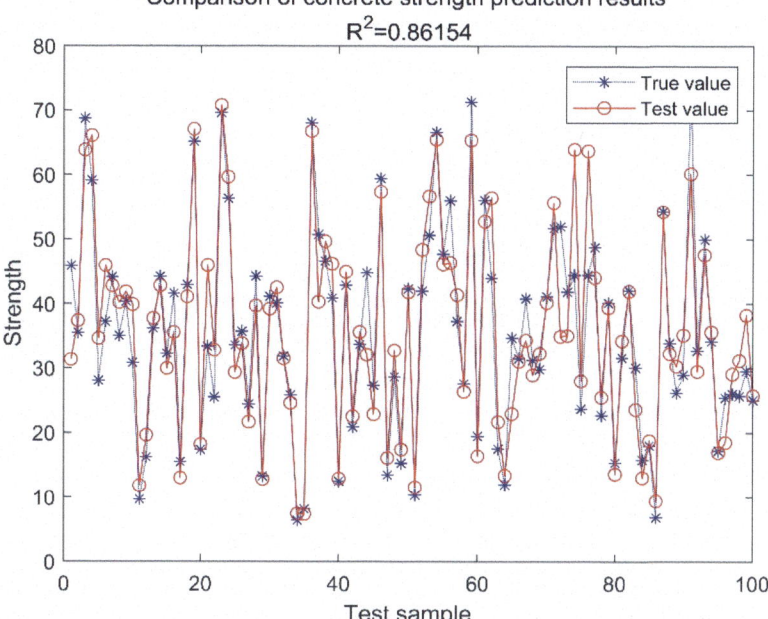

Fig. 4.4 Comparison of concrete strength prediction results

4.2 BP Neural Network Based on Particle Swarm Optimization Algorithm

4.2.1 Overview of Particle Swarm Optimization Algorithm

The particle swarm optimization (PSO) algorithm mimics the foraging behaviour of birds, where individuals initially move randomly but gradually form coordinated groups through information exchange, ultimately converging toward the optimal solution. The food represents the optimal solution required by the neural network [7, 8]. In this analogy, the particles represent factors influencing concrete strength, such as water, cement, and fly ash. These particles move within a fixed range, and although their movements may seem random, they are continuously recording and updating their trajectories to search for the optimal solution in the global search space. The interaction between particles follows an iterative formula [9]:

(1) Update speed: $v_{id}^{k+1} = wv_{id}^k + c_1 r_1 \left(pbest_{id}^k - x_{id}^k\right) + c_2 r_2 \left(gbest_{id}^k - x_{id}^k\right)$
(2) Update location: $x_{id}^{k+1} = x_{id}^k + v_{id}^k + c_1 r_1 (pbest_{id}^k - x_{id}^k) + c_2 r_2 (gbest_{id}^k - x_{id}^k)$

The performance of the particle swarm algorithm is closely related to the selection of parameters. Proper parameter choices are crucial for the convergence speed of PSO-BP and its ability to find the global optimal solution.

PSO is structurally easy to implement, convenient to use, with high accuracy and excellent optimization capabilities. Each particle is closely related to the factors affecting concrete strength, and the algorithm searches for the optimal solution in the global search space. Therefore, PSO is commonly used in research projects and optimization algorithm selections.

4.2.2 PSO-BP Modelling

The role of particle swarm optimization (PSO) in the neural network is to optimize the BP neural network, with parameter settings largely consistent with the initial values of the BP network. Determine the learning factor $c_1 = 1.5$, $c_2 = 1.5$, as they typically range between 1 and 2, generally take 1.5. The inertia weight w decreases linearly between [0.4, 0.8] as the number of iterations increases. The flowchart of the model is shown in Fig. 4.5.

PSO provides an overall optimization of the basic BP neural network, so some parameters are set the same as the initial values in Sect. 4.1.2 for the BP network.

(1) Normalization: The built-in MATLAB function mapminmax is used for standardizing the data, eliminating the influence of different units on the final results;
(2) The number of hidden layers and network layers are set, along with the number of particles in the PSO swarm. Each particle is assigned an initial position and velocity. The initial individual best position (determined by the objective function value) and global best position for each particle are also initialized;
(3) The number of nodes in the input layer is the same as that in the BP network. The neural network's weights and thresholds are randomly initialized, and the fitness values needed for each particle are calculated;
(4) The number of hidden layer nodes is set, and the inertia weight values are updated. The network's performance is treated as the objective function, with the mean squared error (MSE) commonly used as the objective function;
(5) Real-time position and velocity information is updated for the particles to ensure that they are within the limited control area of the model;
(6) Set the output layer to the 28-d compressive strength of ordinary hardened concrete;
(7) Training parameters are set. The algorithm continues to iterate, searching for the optimal solution until the termination criteria are met. If not met, the process returns to step (5).

In PSO, each particle represents an individual in an animal flock, corresponding to different factors influencing concrete strength. The key factors are:

(1) Dimension of solving problem (i.e., number of parameters to be predicted)
(2) Size of the particle swarm (usually depends on prediction region and data samples)

4.2 BP Neural Network Based on Particle Swarm Optimization Algorithm

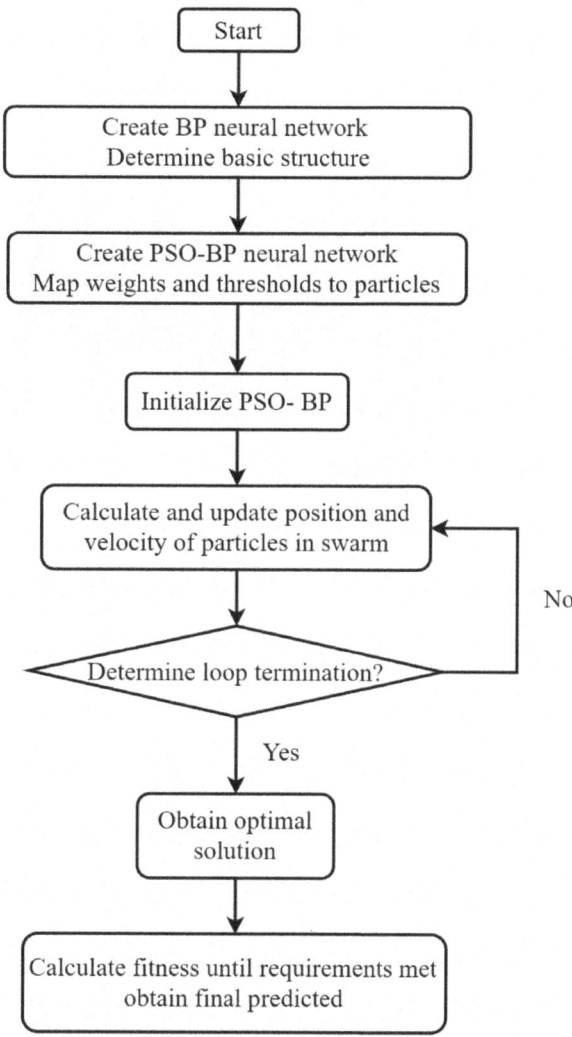

Fig. 4.5 PSO-BP flowchart

(3) Random initialization of each particle's position and velocity (depending on the prediction region, but with a reasonable fluctuation range)

4.2.3 Analysis of Operational Results

The results of the program are shown in Fig. 4.6. It can be seen that the correlation coefficient R^2 of the BP neural network optimized by the particle swarm algorithm (R^2 represents the fitting effect, where a value closer to 1 indicates better fitting) has reached 0.91685. Compared to a single BP neural network, the accuracy has

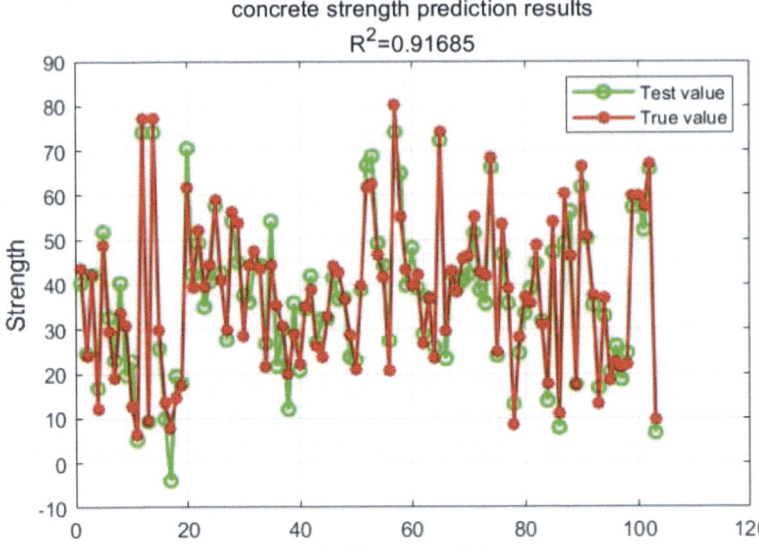

Fig. 4.6 PSO-BP training diagram

significantly improved, making the fitting result closer to 1. Figure 4.7 shows the optimal individual adaptation curve during the PSO-BP optimization process, where the fitness value converged from an initial value of 20 to 17.3.

Figure 4.8 shows the PSO-BP network training diagram, where the components from top to bottom are the number of iterations, iteration time, model performance, model gradient, variable MU, and maximum validation failure count. Figure 4.9 displays the PSO-BP fitting results, including the performance, training state, and regression charts. Analysing the error results of PSO-BP compared to the BP neural network, PSO-BP does not exhibit overfitting, and the values remain stable with minimal fluctuations after each calculation. The precision of the results is also higher, with a smaller relative error range, indicating better accuracy and fitting effect. It is evident that the particle swarm algorithm-optimized BP neural network outperforms the single BP neural network in all aspects, including computational accuracy and handling of discrete data. The training process also demonstrates that the PSO-optimized BP network accelerates when solving complex nonlinear problems, avoiding overfitting and non-convergence issues that could prevent results from meeting expectations.

The fitting coefficients for the test, validation, and overall samples are 0.96416, 0.93269, 0.93779, and 0.95617, respectively, with an average of about 0.95. Due to the increased complexity of the model compared to the single BP network, the result is slightly lower than the above BP fitting coefficients. After performing interval analysis on 1000 data sets at a 95% confidence level, the fitting effect is considered

4.2 BP Neural Network Based on Particle Swarm Optimization Algorithm

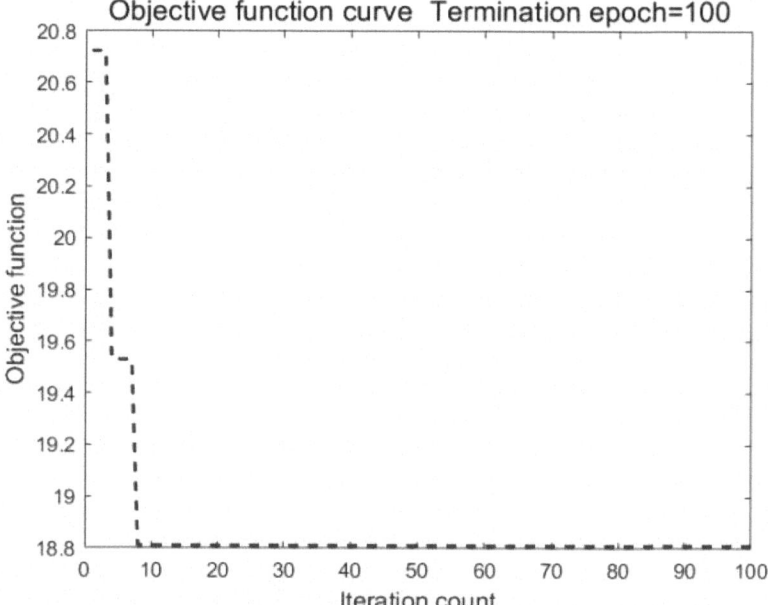

Fig. 4.7 PSO-BP individual adaptation change curve

good. Overall, the accuracy of the results is higher than that of the single BP neural network, meeting the high-precision requirements for engineering applications in predicting hardened concrete strength. Table 4.1

For example, in the test set consisting of 20 groups of data, the data source is from the appendix table. As the absolute error has both positive and negative values, the average in Table 4.1 uses the absolute value. The results show that after PSO optimization, the BP neural network's accuracy has greatly improved. For groups 13 and 14, the data is very close to the measured values. The average absolute error has reached 0.97, nearly a 65% improvement compared to the single BP neural network. This proves that optimizing the single BP neural network with the particle swarm algorithm is reliable. Each particle represents a factor affecting concrete strength, and further expanding and optimizing the data set would significantly increase its practical application value in engineering.

Compared to the error issues of a single BP neural network, PSO-BP conducts a global search for the optimal solution, reducing the project's operation time and avoiding falling into local optima situation.

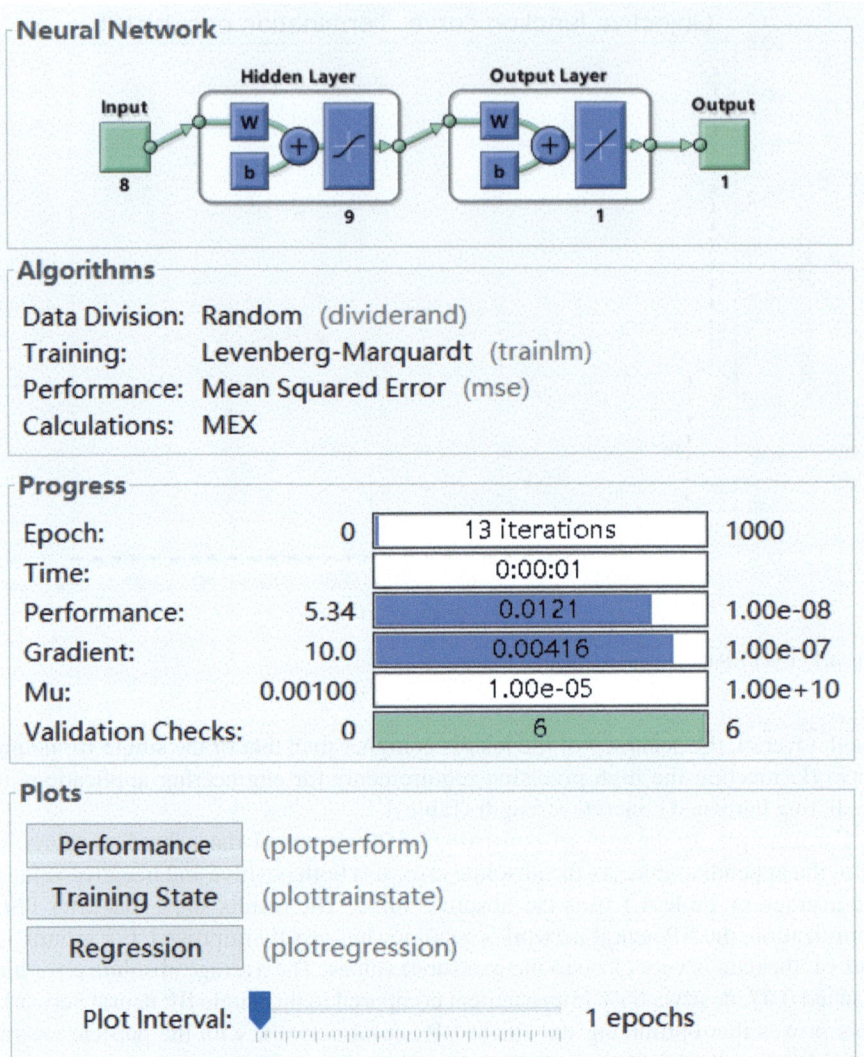

Fig. 4.8 PSO-BP network training diagram

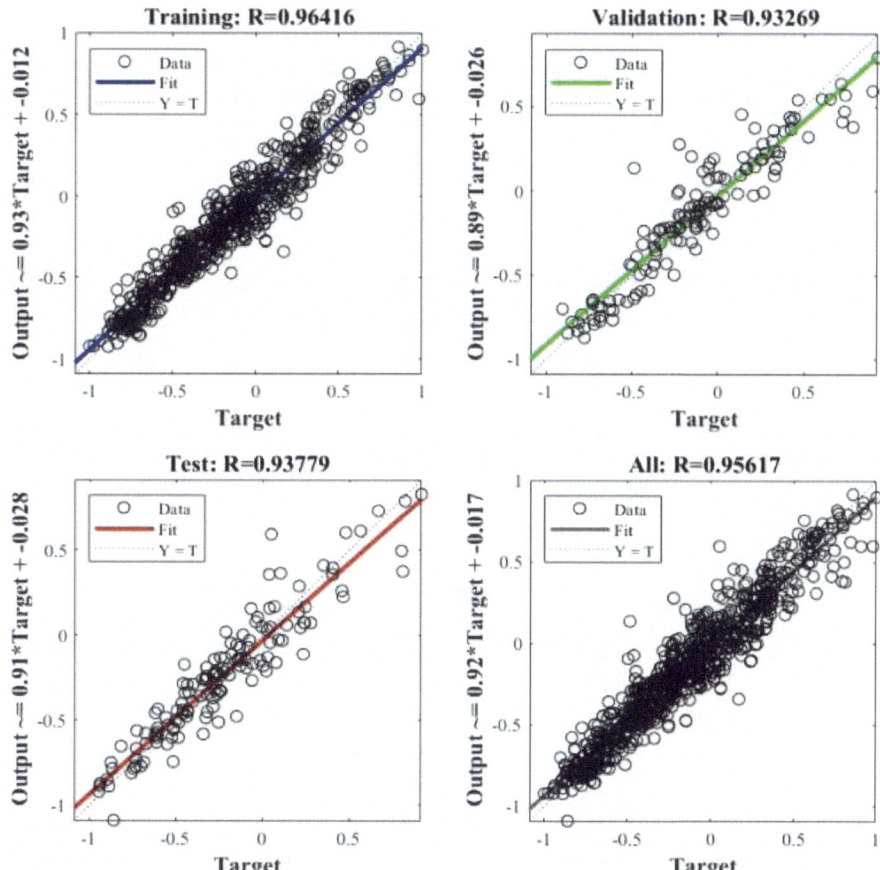

Fig. 4.9 PSO-BP fitting plot

4.3 BP Neural Network Based on Ant Colony Optimization Algorithm

4.3.1 Overview of Ant Colony Optimization Algorithm

Ant Colony Optimization (ACO) is inspired by the foraging behaviour of ants. When searching for food, ants leave pheromones on the paths. Other ants can sense these pheromones and are more likely to follow paths with higher pheromone concentrations, creating a positive feedback loop that gradually leads to the optimal path [10]. This algorithm is unique in its distributed nature, providing timely positive feedback and performing heuristic searches, making it a new heuristic global optimization algorithm in the field of concrete strength prediction.

Table 4.1 Comparison of prediction results before and after PSO optimization

Serial number	True value (MPa)	BP prediction (MPa)	PSO-BP Prediction (MPa)	Absolute error before optimization (MPa)	Absolute error after optimization (MPa)
1	30.2	30.8	29.9	−0.6	0.3
2	22.05	21.4	22.0	0.35	0.05
3	19.81	21.02	20.58	−1.21	−0.77
4	25.5	25.3	25.2	0.2	0.3
5	15.63	19.27	18.96	−3.64	−3.33
6	46.57	30.28	45.28	16.29	1.29
7	35.42	34.20	36.28	1.22	−0.86
8	28.57	20.45	25.63	8.12	2.94
9	57.30	49.23	55.20	8.07	2.1
10	42.32	42.08	43.24	0.24	−0.92
11	10.26	11.23	9.98	−0.97	0.28
12	14.2	13.8	14.1	0.4	0.1
13	17.24	17.19	17.23	0.05	0.01
14	18.96	17.23	19.07	1.73	−0.11
15	20.24	15.99	18.23	4.25	2.01
16	27.34	26.56	27.10	0.78	0.24
17	37.38	38.20	38.06	−0.82	−0.68
18	35.2	32.06	38.08	3.14	−2.88
19	37.23	36.56	37.09	0.67	0.14
20	48.96	47.32	49.10	1.64	−0.14
Average value				2.81	0.97
Maximum value				16.29	3.33

The basic process of the algorithm includes: initializing pheromones, simulating ant behaviour, updating pheromones, calculating path selection probabilities, ant movement, iterative computation, and solution evaluation and update. The process starts with the initialization of pheromones, setting an initial value to simulate the ants' initial preference for path selection. Then, the behaviour of the ants is simulated, including how they explore and choose paths, which simulates the entire search process. Following that, pheromone updating occurs, which is the core of the algorithm. This step influences the future path choices of ants by strengthening or weakening the pheromone concentrations on certain paths.

The calculation of path selection probabilities is based on the current pheromone concentration and the quality of the paths, ensuring that ants tend to choose paths deemed better by previous ants. The ant movement step simulates the actual movement of ants along the selected paths.

4.3 BP Neural Network Based on Ant Colony Optimization Algorithm

Through iterative computation, the algorithm continuously repeats the above process, gradually optimizing the solution to find the optimal one. After each iteration, the solution is evaluated and updated by comparing the new solution with the current best solution, determining whether to update the algorithm's current optimal solution.

The entire process involves complex probability calculations and decision-making, as well as simulating natural ant behaviour and applying the concept of pheromones. These features allow the algorithm to find near-optimal solutions to various optimization problems. Through continuous iteration and updating, the algorithm ultimately converges to the optimal or near-optimal solution, offering an efficient and innovative method for solving complex problems.

4.3.2 ACO-BP Modelling

The role of the ant colony optimization algorithm in the whole neural network is to optimize the BP neural network, and the parameters are set the same as the initial values of the BP neural network. The flow chart is shown in Fig. 4.10.

The ant colony algorithm optimization algorithm serves to optimize the basic BP neural network throughout the whole neural network, and its parameter settings are also similar to the initial value settings of the BP neural network in 4.1.2.

(1) Normalization;
(2) Initialize the ant colony, determine the number of ants and the attribute of each ant, and set the number of network layers;
(3) Set the number of nodes in the input and output layers and initialize the weights and thresholds, again use the local square error as the objective function;
(4) Set the number of nodes in the hidden layer and determine the output layer, each ant represents a solution (a set of weights and thresholds) in the BP neural network, and each ant searches for the combination of weights and thresholds in the solution space according to a certain strategy;
(5) Set the corresponding training parameters to update the pheromone on the ants' path according to the performance of the BP neural network, the pheromone updating rule usually includes pheromone volatilization and pheromone increase;
(6) Calculate path probabilities and make decisions based on the parameters chosen by the ants;
(7) Repeat the ant colony behaviour simulation and pheromone update as well as the path selection probability calculation until a predetermined number of iterations is reached or the termination condition is satisfied.

The ACO algorithm is a search space of weights and thresholds to find the combination of parameters that gives the optimal performance of the neural network, but its performance is also affected by factors such as parameter selection and heuristic information.

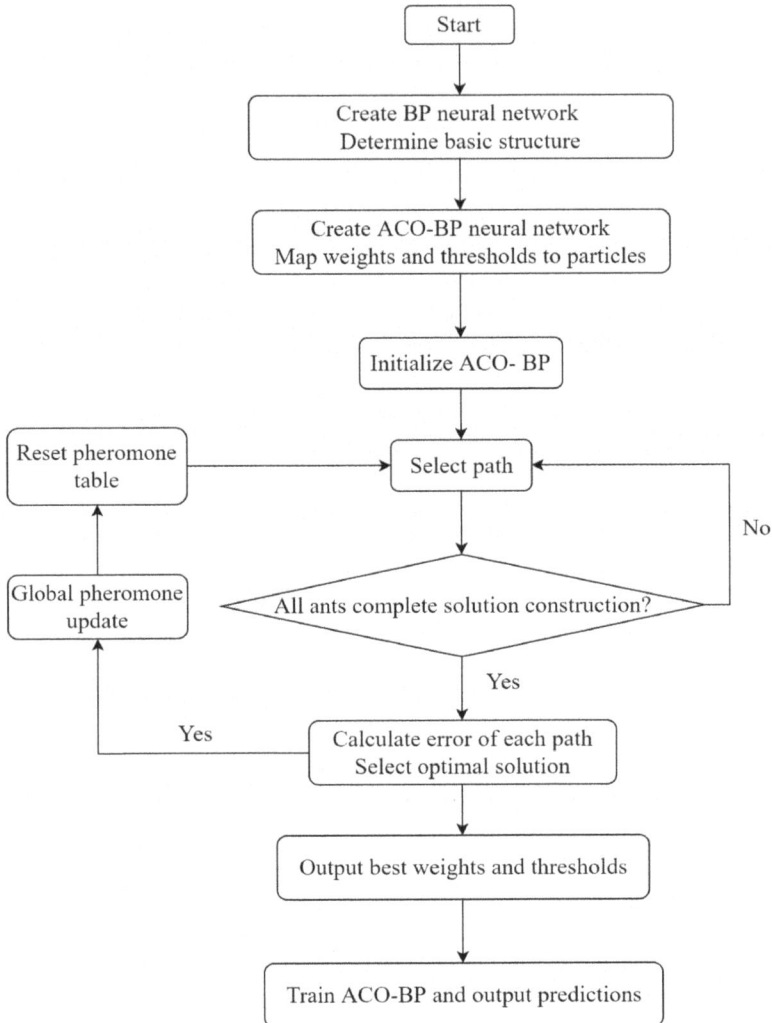

Fig. 4.10 ACO-BP flowchart

4.3.3 Analysis of Operational Results

The program's results are shown in Fig. 4.11. As seen in the figure, after optimization with the ant colony algorithm (ACO), the BP neural network achieves a correlation coefficient R^2 of 0.87658. This represents a significant improvement in accuracy compared to a single BP neural network, though the enhancement is less effective than that achieved using the particle swarm optimization (PSO) algorithm. Figure 4.12 displays the fitness variation curve of the optimal individual during the ACO-BP optimization process, where the fitness value rapidly converges from 20.6 to 18.8.

4.3 BP Neural Network Based on Ant Colony Optimization Algorithm

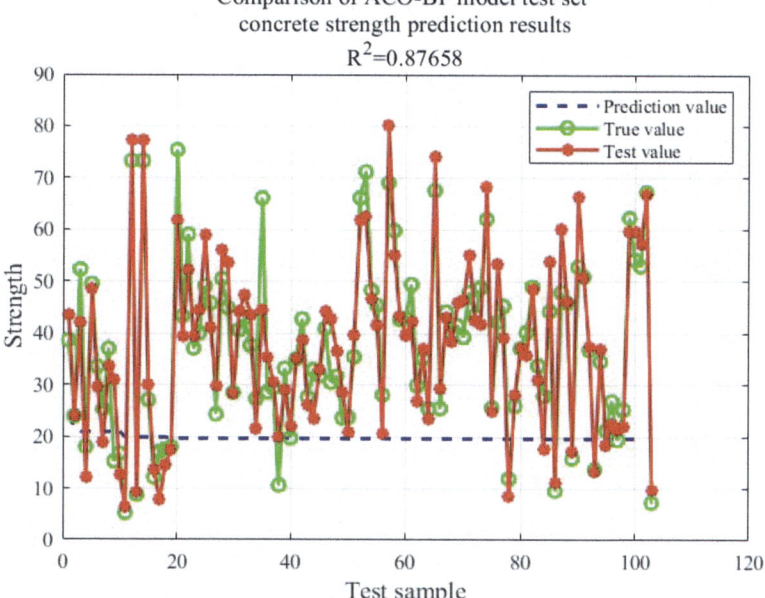

Fig. 4.11 ACO-BP training performance plot

Figure 4.13 illustrates the ACO-BP network training process, while Fig. 4.14 presents the ACO-BP fitting results.

The fitting coefficients for the training, validation, testing, and overall sample sets are 0.95838, 0.93639, 0.94389, and 0.95341, respectively. Although the average value is slightly below 0.95, it still meets accuracy requirements and can be used for data prediction. Compared to the error analysis of the BP neural network, ACO-BP demonstrates greater robustness due to its nature as a parallel algorithm. Without external intervention, it autonomously finds an optimal path from disorder to order. As a positive feedback algorithm, when combined with the BP neural network, both accuracy and fitting performance improve. However, its overall accuracy is slightly lower than that of the PSO-BP model. Table 4.2

A comparison was made using an additional 20 data sets from the test group, sourced from the data set in the appendix table. Considering the issue of sign variations, the average values obtained from Table 4.2 clearly show that the accuracy of the single BP neural network has significantly improved after optimization with ACO, with an overall improvement of approximately 55%. Each ant was used to simulate the mapping of concrete strength factors in the calculation process. However, since the entire calculation requires pheromone updates and iterations, there may be slight deviations in optimization accuracy.

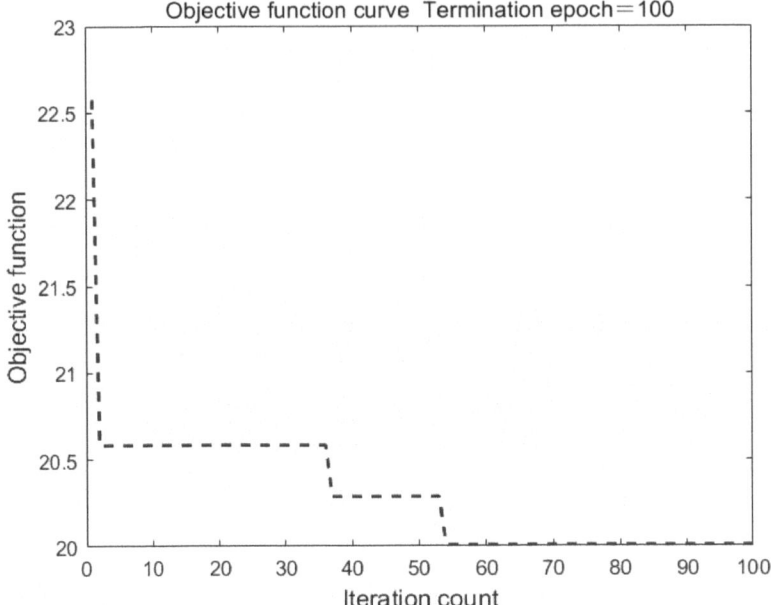

Fig. 4.12 ACO-BP Individual adaptation change curve

4.4 Comparison of Model Prediction Performance

The results of the performance comparison between PSO-BP model and ACO-BP model are shown in Fig. 4.15.

Both optimization algorithms offer positive improvements for the single BP neural network. However, as shown in Fig. 4.15, a comparison between the PSO-BP and ACO-BP models reveals that the PSO-BP model has a smaller relative error compared to ACO-BP. Additionally, as the database is updated, to study early concrete strength, data sets with excessive age must be discarded. In this case, the advantages of the PSO-BP model become more apparent. Compared to the self-organizing learning process of the ACO algorithm, the PSO algorithm has faster search speeds, higher efficiency, and a more advantageous algorithm structure. From Table 4.3, the correlation coefficient between BP neural network and ACO-BP only increases by 0.01595 (approximately 1.8%), from 0.86063 for BP to 0.87658 for ACO-BP. In contrast, the improvement from BP to PSO-BP is 0.5622 (approximately 6.5%), clearly indicating that PSO-BP has superior accuracy to ACO-BP.

In methods for predicting early concrete strength, the PSO algorithm, by optimizing the weights and biases of the BP neural network, can more accurately map the complex nonlinear relationship between the concrete's basic parameters and its

4.4 Comparison of Model Prediction Performance

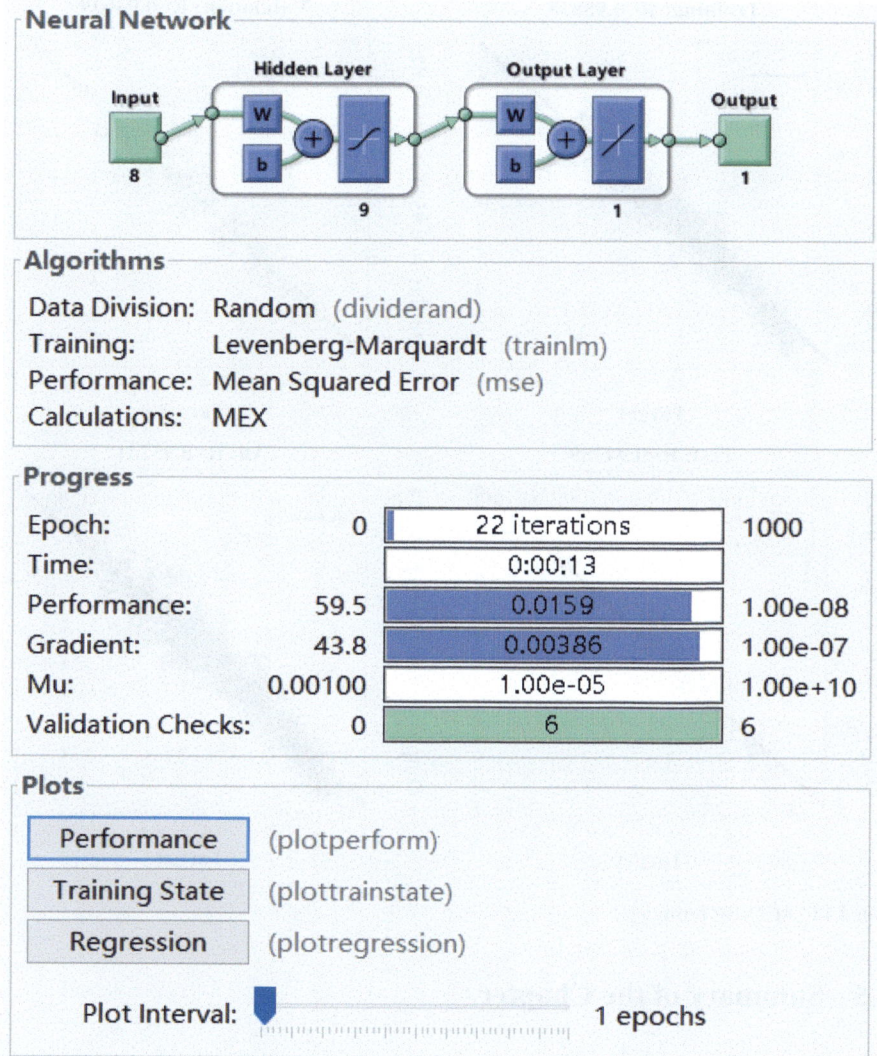

Fig. 4.13 ACO-BP network training diagram

strength. PSO's global search capability effectively avoids local optima in multidimensional search spaces, thus enhancing the accuracy and reliability of predictions. This advanced search strategy makes the PSO-BP neural network model especially suitable for predicting the performance of materials like concrete, providing a powerful tool for early strength prediction based on various influencing factors.

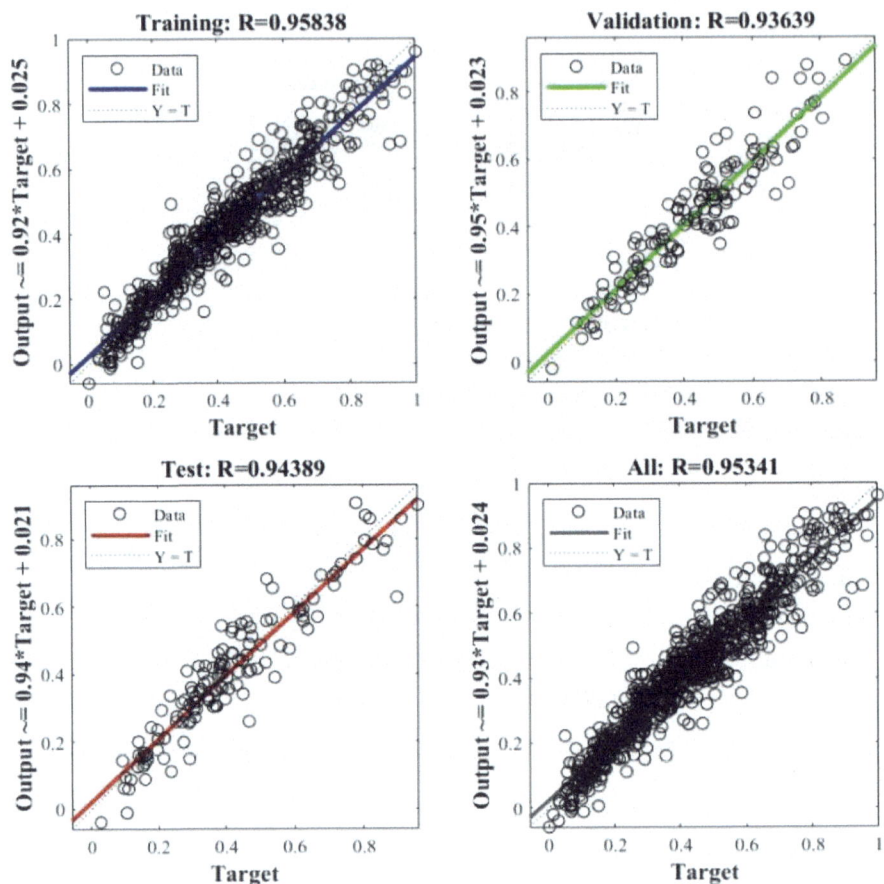

Fig. 4.14 ACO-BP fitting plot

4.5 Summary of the Chapter

This author compares the performance of two optimized neural network models with the original BP neural network model to select the most suitable model for subsequent development of an intelligent prediction program. In the validation set of 100 selected groups, 20 randomly chosen validation data sets revealed that the average absolute error of the PSO-BP model is 0.97, while that of the ACO-BP model is 1.49, indicating that the optimized PSO-BP performs better than the ACO-BP.

Based on the data above, the following conclusions can be drawn:

(1) The ant colony optimization (ACO) algorithm simulates the path selection mechanism of ants searching for food, using collective cooperation to find the optimal solution. The particle swarm optimization (PSO) algorithm simulates the social behaviour of bird flocks, achieving optimization through information sharing

Table 4.2 Comparison of predictions before and after ACO optimization

Serial number	True value (MPa)	BP prediction (MPa)	ACO-BP prediction (MPa)	Absolute error before optimization (MPa)	Absolute error after optimization (MPa)
1	43.24	40.24	41.10	3.0	2.14
2	23.98	27.34	20.98	−3.36	3.0
3	56.37	54.23	55.20	2.14	1.17
4	26.37	27.89	25.10	−1.52	1.27
5	19.87	13.57	24.30	6.3	−4.43
6	26.30	27.30	26.39	−1.0	−0.09
7	43.29	42.32	45.23	0.97	−1.94
8	30.98	27.30	31.20	3.68	−0.22
9	51.20	44.20	52.98	7.0	−1.78
10	62.10	59.58	61.09	2.52	1.01
11	8.07	10.20	9.02	−2.13	−0.95
12	13.20	12.98	13.28	0.22	−0.08
13	18.20	10.89	20.0	7.31	−1.8
14	26.39	25.39	27.20	1.0	−0.81
15	17.1	18.20	17.23	−1.1	−0.13
16	30.2	25.39	27.30	4.81	2.9
17	27.38	26.30	28.10	1.08	0.72
18	44.32	42.39	43.20	1.93	1.12
19	20.19	15.20	17.99	4.99	2.2
20	50.21	39.20	48.20	11.01	2.01
Average value				3.35	1.49
Maximum value				11.01	4.43

among particles. Both methods enhance the learning ability of the BP neural network, effectively preventing the network from getting stuck in local optima during training, thus improving the accuracy and reliability of the prediction model.

(2) The PSO algorithm shows a more significant advantage over ACO in optimizing the weights and thresholds of the BP neural network. By simulating the social behaviour of fish or birds, PSO enables individual particles to search for the optimal solution in the solution space while continuously adjusting their flight direction and position. This mechanism gives PSO a stronger global search capability, effectively avoiding local optima. Additionally, PSO can more accurately model the influence of various factors on concrete strength, further improving prediction accuracy and fitting performance, thus demonstrating superior performance in predicting concrete strength.

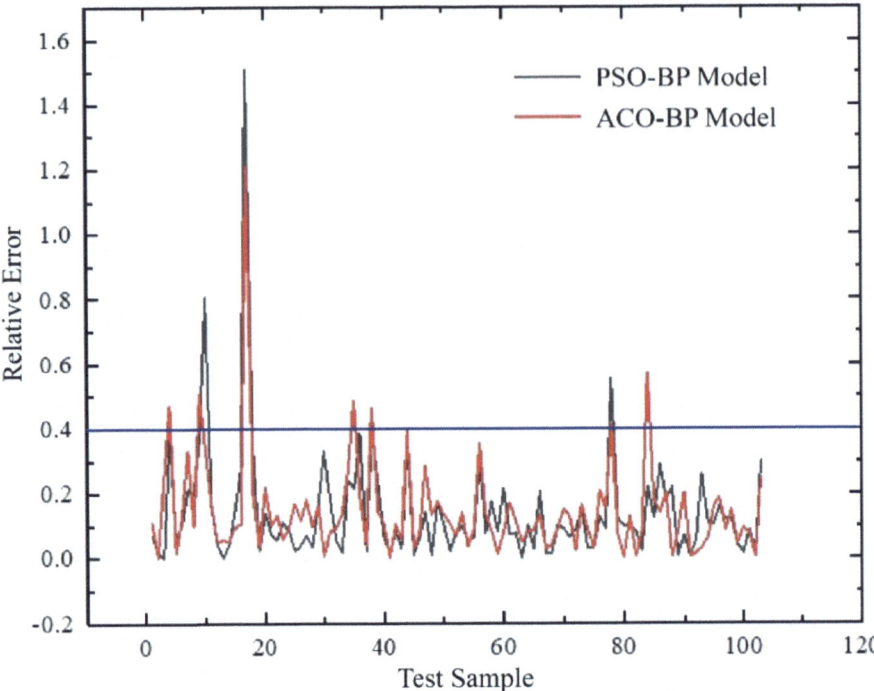

Fig. 4.15 Comparison of optimization models

Table 4.3 Comparison of correlation coefficients

–	BP	ACO-BP	PSO-BP
R^2	0.86063	0.87658	0.91685

(3) Using the PSO-BP model as the core of the intelligent prediction program, combined with the traditional strength prediction method of the F-P maturity equation, a complete concrete strength prediction system is established. This combination leverages the powerful ability of the PSO-BP neural network in handling complex, nonlinear problems and the effectiveness of the F-P maturity equation in early concrete strength prediction, providing a more comprehensive and accurate prediction tool. This integrated method not only expands the applicability of the prediction model but also enables more precise predictions of concrete strength changes under different conditions, offering strong technical support for the design and construction of concrete structures, optimizing resource allocation, and improving engineering efficiency and safety.

References

1. Lantz Brett. Machine learning and R language [M]. Machine Learning and R language, 2015.
2. Chen HG, Long WY, Li X et al (2021) Prediction of compressive strength of fly ash concrete with BP neural network[J]. Building Structure S02:51–54
3. Luo B, Huang W J, Yang S. Improved algorithm of BP neural network based on parameters adjustment of Tan-Sigmoid transfer function [J]. Journal of Chongqing University (Natural Sciences Edition), -2006(01): 150–153.
4. Gao DW, Wang P, Cai ZC (2003) Optimization of hidden nodes and training times in artificial neural network [J]. Journal of Harbin Institute of Technology. 35(2):207–209
5. Peng Z B, Gao Y. Application of BP neural network in elevation fitting of underwater terrain [J]. Journal of Chongqing Jiaotong University (Natural Science Edition), -2018, 37(11): 64–68, 82.
6. Yuan D B, Zhang J, Zhao C W, et al. GNSS elevation fitting based on improved RBF neural network[J]. Geodesy and Geodynamics, -2020, 40(3): 221–224, 241.
7. Jain M, Saihjpal V, Singh N et al (2022) An Overview of Variants and Advancements of PSO Algorithm[J]. Appl Sci 12(17):8392
8. Marini F, Walczak B. Particle swarm optimisation (PSO). A tutorial[J]. Chemometrics and Intelligent Laboratory Systems. 2015, 149: 153–165.
9. Zhu M T, Liu H, Wu S Y, et al. Particle swarm algorithm with hybrid strategy improvement[J]. Journal of Chongqing University of Technology (Natural Science), -2024, 38(01): 110–121.
10. Kumar S, Kumar-Solanki V, Choudhary S K, et al. Comparative Study on Ant Colony Optimization (ACO) and K-Means Clustering Approaches for Jobs Scheduling and Energy Optimisation Model in Internet of Things (IoT)[J]. International journal of interactive multimedia and artificial intelligence,- International Journal of the Internet of Things -. 2020, 6(1): 107–116.

Open Access This chapter is licensed under the terms of the Creative Commons Attribution-NonCommercial-NoDerivatives 4.0 International License (http://creativecommons.org/licenses/by-nc-nd/4.0/), which permits any noncommercial use, sharing, distribution and reproduction in any medium or format, as long as you give appropriate credit to the original author(s) and the source, provide a link to the Creative Commons license and indicate if you modified the licensed material. You do not have permission under this license to share adapted material derived from this chapter or parts of it.

The images or other third party material in this chapter are included in the chapter's Creative Commons license, unless indicated otherwise in a credit line to the material. If material is not included in the chapter's Creative Commons license and your intended use is not permitted by statutory regulation or exceeds the permitted use, you will need to obtain permission directly from the copyright holder.

Chapter 5
Development of Intelligent Concrete Strength Prediction Program

This chapter describes the development of an intelligent concrete strength prediction program using MATLAB software, based on the models established in Chaps. 3 and 4. The program's functionality and interface were optimized to ensure ease of use and maintainability for future updates. The development process of the intelligent program in MATLAB is detailed, demonstrating how the program can predict concrete strength on-site by using known input and output parameters. The prediction results are then compared with actual values from mix proportion data, validating the accuracy and reliability of the concrete compressive strength prediction program.

5.1 Analysis and Implementation of Intelligent Concrete Strength Prediction Program

5.1.1 Theoretical Foundations of MATLAB Program Development

When developing the intelligent program, we considered its inputs and outputs, similar to the relationship between main functions and sub-functions. For ease of use in concrete strength prediction, the program must be highly maintainable and capable of database updates. Therefore, MATLAB's built-in visual interface editor is used to create a visual input/output interface. This allows for visual input of the eight different parameters and visual output of the prediction results, providing significant convenience for practical engineering applications.

As an advanced programming language with its own interactive environment, MATLAB has become increasingly popular for developing intelligent prediction programs. The software's built-in graphical user interface (GUI) functions allow for designing the app using various elements such as buttons, menus, and text boxes.

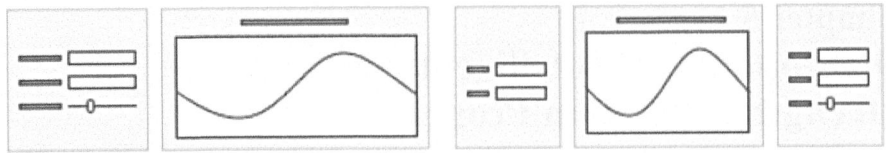

Fig. 5.1 APP designer user interface

MATLAB's App Designer is a visual development environment used to create GUIs. It enables users to design the desired program interface by dragging and dropping components and using the interactive editor, then adding the necessary code to ensure proper functionality. Figure 5.1 shows the user interface of MATLAB's App Designer, which includes the UIFigure and Component Browser. Components are selected and dragged into the UIFigure, and the Component Browser displays the components currently added. Users can modify component names in the browser, and the changes are automatically reflected in the code.

This study combines the concrete strength prediction methods from Chaps. 3 and 4 with MATLAB code to develop the intelligent prediction program, fully utilizing MATLAB's toolboxes and built-in functions.

5.1.2 Intelligent Program Function Module Design

When developing the basic intelligent program, modular programming must first be implemented. In MATLAB, related functions are defined using the 'function' keyword. A .m file typically contains the basic definition of one function. The basic MATLAB development environment (IDE) easily supports syntax highlighting and code indentation, allowing the function to be saved independently.

After defining the function, it needs to be called. If the function is used in other scripts or functions, it can be called by its name and parameters. It's important to distinguish between local variables, as some variables are only visible within the function and cannot pollute the global scope. Similarly, the 'global' command can be used to adjust variables and make them global.

The program requires a database, and function files also need documentation to support. The documentation should describe the function's purpose, input parameters, output parameters, and methods used. The 'help' command in MATLAB can be used to view this documentation.

The 'varargin' and 'varargout' commands allow for variable numbers of input and output parameters. In the program written in this study, there are eight input parameters and one output parameter. A key issue is the function handle. The two different models for predicting concrete strength, described in Chaps. 3 and 4, should exist as separate functions. Both functions coexist but have different definitions, so the function should be passed as a parameter to other functions to ensure that new functions are dynamically created during runtime.

For the design of the intelligent program's functional modules, MATLAB requires unit tests to verify the correctness of the functions, ensuring that they work correctly under different environments. Additionally, MATLAB should use vectorized operations to improve code execution efficiency.

5.1.3 Intelligent Program Design Process

The design process of developing an intelligent program using MATLAB is as follows:

1. Open MATLAB APP Designer by typing 'app designer' in the MATLAB command window and pressing the Enter key, or by selecting App Designer under the APPS tab from the main MATLAB interface;
2. Create a new App by selecting 'Blank App' to start a new project. The app design interface consists of a component library, UIfigure, and component browser. The component library is used for page design, with different libraries offering different functionalities. The UIfigure is the container and includes both design and code views;
3. Design the user interface by finding various UI components, such as buttons, text boxes, and drawing areas, in the left toolbox of App Designer. Drag these components onto the interface, adjusting their size and position as needed;
4. Write MATLAB code, select the Code View tab, where you write the MATLAB code that interacts with the interface, add callback functions for each component to respond to user actions, this can be done by setting up callback functions for each component in the Property Inspector to ensure the proper operation of the software.
5. During the design phase, click the 'Run' button to test the app, allowing you to detect and resolve any code errors or issues. Add detailed comments to function files to explain the purpose of each step and the inputs and outputs of functions, which helps with understanding and use;
6. When satisfied with the app and ready to share it, choose the preferred deployment method. Since not all computers have MATLAB installed, MATLAB Compiler can be used to generate an executable file along with MATLAB Runtime. The corresponding version of MATLAB Runtime must be installed on the machine, allowing the MATLAB program to run even without MATLAB;
7. Continuously update and maintain the database and app, improving and optimizing based on user feedback and requirements.

5.2 Intelligent Concrete Strength Prediction Program Design

The concrete strength prediction methods discussed in Chaps. 3 and 4—he F-P maturity equation-based model and the PSO-BP artificial neural network model—are both important technologies in the field of concrete strength prediction, each with its own unique application scenarios and advantages. With the widespread use of MATLAB, the similarities in the model development process for these two methods have become more apparent, as both rely on known input parameters (such as concrete mix proportion, water–binder ratio, age, etc.) and corresponding output data (concrete strength) for training and fitting.

The F-P maturity equation is a prediction model based on the cumulative effects of concrete age and temperature, suitable for predicting early concrete strength development. Its advantage lies in its simplicity, making it easy to understand and apply, especially when data is limited or when the primary prediction focuses on early strength. On the other hand, the PSO-BP artificial neural network-based method adopts the PSO algorithm to optimize the weights and biases of a back propagation (BP) neural network to improve prediction accuracy. This method can handle nonlinear and complex relationships and is particularly effective with large datasets, especially when relationships between variables are unclear or when there is high nonlinearity in the data.

Compared to the F-P maturity equation, the artificial neural network offers greater flexibility, as it does not require an explicit definition of the model's form. It can analyse databases, learning complex patterns and relationships. Therefore, by combining both methods, a more complex nonlinear relationship is established, and an intelligent program can be created using MATLAB for rapid predictions on construction sites. Figure 5.2 shows the design flowchart for the intelligent concrete compressive strength prediction program.

Based on the above flowchart, the two models' code is combined.

The integration of the two methods is not simply about listing the code for calculation but about improving the final program's accuracy through ensemble learning. This is achieved by merging models in various strategies, using stacking as an example, combining the PSO-BP neural network model and the F-P maturity equation model.

1. Using the selected dataset (see Appendix A), the Stacking process is divided into two stages: the first is the training stage, and the second is training the meta-model to combine the two models above.
2. The same type of training and test datasets as in Chaps. 3 and 4 are used to build the models.
3. The base models are created and presented in the form of code for the neural network and F-P maturity equation, then calculations are conducted. The stacking model uses the predictions from the previous models as inputs to train a meta-model. Figure 5.3 displays the schematic of the Stacking model.

5.2 Intelligent Concrete Strength Prediction Program Design

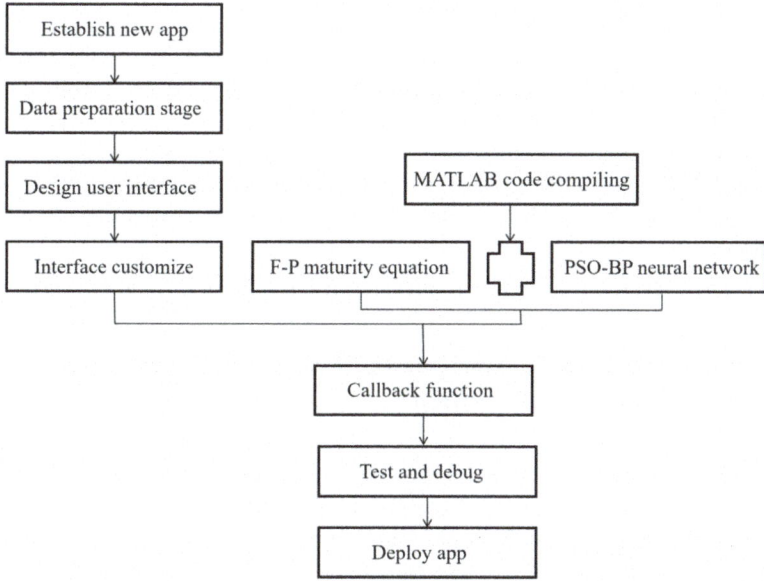

Fig. 5.2 Flow chart of intelligent prediction procedure design for concrete compressive strength

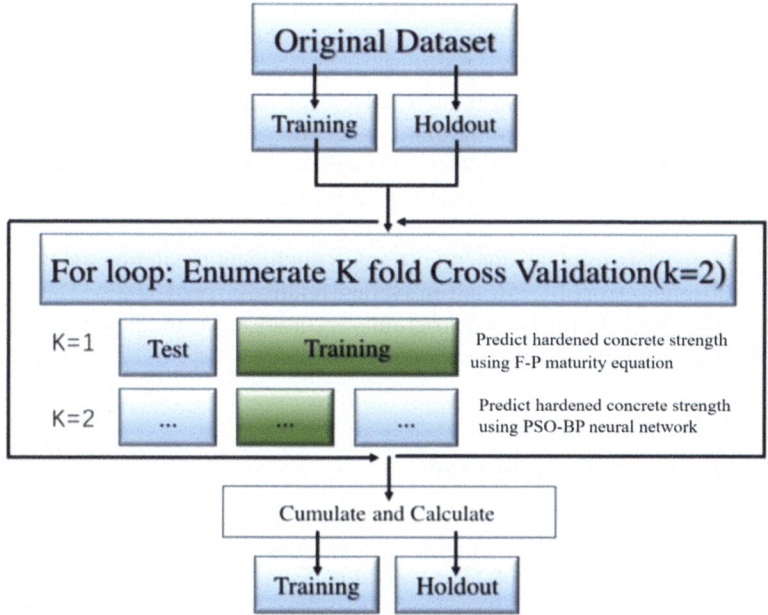

Fig. 5.3 Schematic diagram of stacking model

After writing and saving the MATLAB program, to ensure it runs independently without relying on MATLAB's complete environment and interface, MATLAB Runtime must be installed. MATLAB Runtime is a collection of shared libraries, MATLAB code, and other files that allow compiled applications to run on systems without MATLAB, and it is compatible with various operating systems. Then, using the MATLAB Compiler SDK, the two components can be packaged, allowing the intelligent program to run on different computers without requiring MATLAB installation.

5.3 Concrete Strength Intelligent Prediction Program Performance Measurement

This software primarily adopts databases from literature and books. To verify the accuracy and performance of the software, data from different types of concrete mix proportions provided by a batching plant in Ningbo were used. Table 5.1 shows the batching plant's mix proportion data, while Table 5.2 compares the predicted values and actual strength results.

The 22 mix proportions data adopted—are all provided by the Ningbo batching plant. To account for potential sign differences, the absolute errors were listed in the table for easier reference. From the table, it can be seen that only the fifth and twenty-second sets of data have relative errors exceeding 10%, while the rest have relative errors within 10%. The relative errors for the seventh and eleventh sets are as low as 0.14%, demonstrating that the accuracy of this intelligent prediction program is suitable for practical engineering applications. To provide a more intuitive understanding of prediction accuracy, the relative errors are plotted as a bar chart in Fig. 5.4.

The predicted data from the concrete strength intelligent prediction program have an average absolute error of 1.756 and an average relative error of 3.92%. Therefore, it can be concluded that the prediction program's accuracy meets the standards required for practical engineering applications.

5.4 Summary of the Chapter

This chapter aims to establish an intelligent prediction system for concrete compressive strength by using the models developed in Chaps. 3 and 4. After importing the collected database into MATLAB software, the entire software development process

5.4 Summary of the Chapter

Table 5.1 Mixing plant mixing proportion data

Serial number	Strength grade	Water (kg)	Cement (kg)	Fly ash (kg)	Blast furnace slag (kg)	Water reducer (kg)	Coarse aggregate (kg)	Fine aggregate (kg)
1	C30	160	205	86	82	3.72	1005	789
2	C30	163	204	93	74	3.7	1004	789
3	C30	160	227	63	82	3.72	1023	772
4	C35	160	234	70	86	3.9	1036	750
5	C35	163	221	70	97	3.88	1035	750
6	C35P6	160	264	56	80	4	1055	732
7	C35P6	160	280	68	52	4	1042	754
8	C35	160	264	56	80	4	1054	732
9	C40	160	286	63	72	4.21	1041	754
10	C30	160	240	88	72	4	1018	768
11	C35P10	160	280	68	52	4	1042	754
12	C35P10	160	270	90	50	4.1	1035	750
13	C40	160	303	63	55	4.21	1041	754
14	C40	160	286	88	66	5.35	924	809
15	C45	158	307	43	77	5.12	1065	740
16	C50	153	320	90	86	5.4	1069	711
17	C50	153	320	45	115	5.4	1042	695
18	C50	155	324	60	115	5.99	924	790
19	C60	156	405	58	116	6.59	896	765
20	C35P8	160	300	56	44	4	1035	750
21	C35	160	264	72	64	4	1054	732
22	C35	160	260	80	60	4	1054	732

was defined. By examining the application of MATLAB software in various literature sources, this study aligns with the new developments and changes in the intelligent era. It focuses on using MATLAB to develop an intelligent prediction system for compressive strength, aiming to provide rapid predictions on construction sites, reduce risks, improve quality control, and optimize the construction process. The conclusions of this chapter are as follows:

1. Based on the mix proportion data provided by the Ningbo batching plant, the accuracy of the compressive strength intelligent prediction program was tested. A comparison between the actual strengths from the 22 mix proportions and the predicted strengths from the intelligent program shows that only two sets of data had relative errors above 10%. This proves that the accuracy of the prediction program meets the standards required for practical engineering.

Table 5.2 Comparison of true and predicted intensity results

Serial number	True value (MPa)	Predicted value (MPa)	Absolute error (MPa)	Relative error (%)
1	34.1	32.9641	1.1359	3.33
2	33.1	32.2808	0.8192	2.47
3	34.5	35.5674	1.0674	3.09
4	43.6	42.5673	1.0327	2.37
5	39.8	34.1733	5.6267	14.13
6	42.4	39.9765	2.4235	5.72
7	42.8	42.741	0.059	0.14
8	45.4	46.8762	1.4762	3.25
9	46.6	47.351	0.751	1.61
10	35.6	38.2992	2.6992	7.58
11	42.8	42.741	0.059	0.14
12	38.9	39.5019	0.6019	0.15
13	46.4	48.624	2.224	4.79
14	45.2	46.2945	1.0945	2.42
15	50.7	49.853	0.847	1.67
16	57.8	55.2998	2.5002	4.33
17	59.5	57.0759	2.4241	4.07
18	58.5	60.8048	2.3048	3.94
19	64.5	66.1288	1.6288	2.53
20	43.5	43.6808	0.1808	0.42
21	42.8	40.5131	2.2869	5.34
22	41.9	36.5189	5.3811	12.80

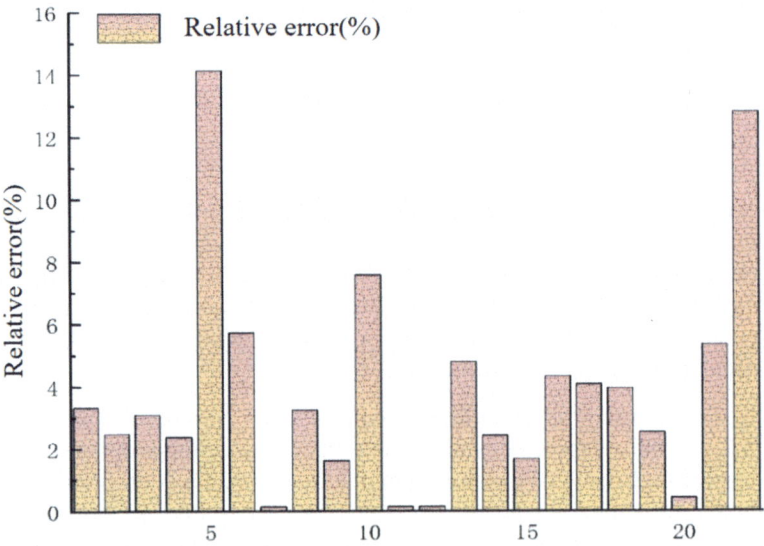

Fig. 5.4 Comparison of relative error

5.4 Summary of the Chapter

2. The intelligent prediction program's code consists of two different concrete strength prediction methods. The program's accuracy confirms that these two prediction models are suitable for calculating the compressive strength of ordinary hardened concrete. The complementary nature of the two methods provides a more comprehensive safety foundation for future engineering projects, improve productivity, and help construction units effectively schedule construction plans. This reduces waiting time for concrete hardening and minimizes construction costs.

Open Access This chapter is licensed under the terms of the Creative Commons Attribution-NonCommercial-NoDerivatives 4.0 International License (http://creativecommons.org/licenses/by-nc-nd/4.0/), which permits any noncommercial use, sharing, distribution and reproduction in any medium or format, as long as you give appropriate credit to the original author(s) and the source, provide a link to the Creative Commons license and indicate if you modified the licensed material. You do not have permission under this license to share adapted material derived from this chapter or parts of it.

The images or other third party material in this chapter are included in the chapter's Creative Commons license, unless indicated otherwise in a credit line to the material. If material is not included in the chapter's Creative Commons license and your intended use is not permitted by statutory regulation or exceeds the permitted use, you will need to obtain permission directly from the copyright holder.

Chapter 6
Conclusions and Foresight

6.1 Conclusions of the Study

As artificial intelligence technology increasingly integrates with the construction industry, the rapid and convenient prediction of concrete strength has become a critical need in engineering practice. Relying on traditional methods of concrete strength prediction is not only inefficient but also challenging to ensure accuracy, and any omissions or errors can lead to severe consequences. By combining the maturity equation and artificial neural networks, and implementing them using MATLAB software, this study aims to enhance the predictive capability of concrete strength under various environmental and mix proportion conditions. After training with external data, the effectiveness and practicality of this intelligent prediction system were successfully validated using mix proportion data from the Ningbo batching plant. The main research findings of the present study are as follows:

Firstly, this study thoroughly analysed the core concepts and calculation methods of the F-P maturity equation, calculating the reaction rate constants and apparent activation energy for different temperature ranges (10–20 °C and 20–35 °C). Based on these parameters, equivalent age was calculated, and the compressive strength of concrete was predicted using the maturity equation. To optimize the calculation process, related steps have been integrated into the program code to improve operational efficiency. This research provides an effective methodological basis for predicting concrete compressive strength.

In addition, this study delves into the basic structure and key concepts of artificial neural networks, starting with a single BP neural network and elaborating on its establishment and parameter tuning process. Subsequently, the BP neural network was optimized using Particle Swarm Optimization (PSO) and Ant Colony Optimization (ACO), detailing the optimization strategies and tuning methods of both algorithms. After comparison, it was determined that the PSO-optimized artificial neural network (PSO-BP) outperformed the ACO-optimized network (ACO-BP) in terms of accuracy and fitting speed in predicting concrete strength. Therefore, the

PSO-BP model was integrated into the intelligent prediction system to enhance the performance of the intelligent program.

Thirdly, this study successfully developed an intelligent program for predicting concrete compressive strength, combining the F-P maturity equation model and the PSO-optimized BP neural network (PSO-BP) model. By coding these two prediction models in the MATLAB environment and using a database for training and self-updating, this program can predict concrete strength efficiently and accurately. Further, by validating with specific mix proportion data from the Ningbo batching plant, the practicality and reliability of this intelligent prediction program were confirmed. To facilitate widespread use, the program supports operation through the installation of MATLAB Runtime, allowing the intelligent prediction tool to be used without installing MATLAB software.

6.2 Innovation Points

In this study, the F-P maturity model is codified by simplifying the complex steps in the model and gradually integrated into the final intelligent prediction program. This process avoids complex conversion steps, enhancing both the efficiency and accuracy of the prediction model.

In this study, the BP artificial neural network is optimized by applying particle swarm algorithm (PSO) and ant colony algorithm (ACO) and the performance of the two methods after optimization is comparatively analysed. Using only basic parameters of concrete, the BP neural network optimized with PSO achieves more accurate prediction strength.

The F-P maturity model is combined with the PSO-BP artificial neural network model, leveraging their dependency on historical and experimental data to fit nonlinear relationships through parameter and weight optimization. Using MATLAB software, these two models are integrated in code form to identify key influencing factors they might share at the output end. This method aims to achieve precise strength predictions to support rapid prediction needs at construction sites.

6.3 Future Prospects

During the construction of the F-P maturity equation, the effects under negative temperature conditions were not considered. Similarly, when improving the F-P maturity equation later, the corrections made using the equivalent age and temperature fitting relationship also did not account for the impact of negative temperatures. Therefore, future research should further refine the F-P maturity equation, fully considering the influence of different environmental conditions, especially negative temperatures, to enhance the model's accuracy and applicability.

6.3 Future Prospects

To further improve the performance of artificial neural networks in predicting concrete strength, it is recommended to conduct in-depth studies and analyses of existing literature to explore how different optimization algorithms can effectively improve the model's prediction accuracy. By comparing the advantages and limitations of various optimization algorithms, the most suitable ones for concrete strength prediction can be identified. Additionally, exploring strategies for combining different models, such as ensemble learning or multi-model fusion, may further enhance the robustness and generalization of the model. Therefore, proposing and validating diverse model combinations and their application in concrete strength prediction will be an important direction for future research. This will not only improve model performance but also provide more precise theoretical support for concrete material quality control and optimization.

To enhance the practicality and portability of the concrete compressive strength intelligent prediction program, the development focus should be on breaking the dependency on MATLAB Runtime, making it an independent application. This means that the developed intelligent prediction program should be able to run independently on various computing platforms, such as personal computers, workstations, and even mobile devices, without requiring a MATLAB environment. Using a cross-platform development framework and refactoring the core algorithms in a general-purpose programming language, we can ensure the program's broad availability and ease of deployment.

Open Access This chapter is licensed under the terms of the Creative Commons Attribution-NonCommercial-NoDerivatives 4.0 International License (http://creativecommons.org/licenses/by-nc-nd/4.0/), which permits any noncommercial use, sharing, distribution and reproduction in any medium or format, as long as you give appropriate credit to the original author(s) and the source, provide a link to the Creative Commons license and indicate if you modified the licensed material. You do not have permission under this license to share adapted material derived from this chapter or parts of it.

The images or other third party material in this chapter are included in the chapter's Creative Commons license, unless indicated otherwise in a credit line to the material. If material is not included in the chapter's Creative Commons license and your intended use is not permitted by statutory regulation or exceeds the permitted use, you will need to obtain permission directly from the copyright holder.

Appendix
Summary of Hardened Concrete Compressive Test Data Used in This Study

Cement (kg)	Blast furnace slag (mining) (kg)	Fly ash (kg)	Water (kg)	Water reducer (kg)	Coarse aggregate (kg)	Fine aggregate (kg)	Age (d)	Compressive strength (MPa)
540	0	0	162	2.5	1040	676	28	79.99
540	0	0	162	2.5	1055	676	28	61.89
332.5	142.5	0	228	0	932	594	270	43.24
332.5	142.5	0	228	0	932	594	365	41.05
198.6	132.4	0	192	0	978.4	825.5	360	44.3
266	114	0	228	0	932	670	90	47.03
380	95	0	228	0	932	594	365	43.7
380	95	0	228	0	932	594	28	36.45
266	114	0	228	0	932	670	28	45.85
475	0	0	228	0	932	594	28	39.29
198.6	132.4	0	192	0	978.4	825.5	90	38.07
198.6	132.4	0	192	0	978.4	825.5	28	23.98
427.5	47.5	0	228	0	932	594	270	43.01
190	190	0	228	0	932	670	90	42.33
304	76	0	228	0	932	670	28	47.81
380	0	0	228	0	932	670	90	56.37
139.6	209.4	0	192	0	1047	806.9	90	39.36
342	38	0	228	0	932	670	365	56.14
380	95	0	228	0	932	594	90	40.56
475	0	0	228	0	932	594	180	42.62
427.5	47.5	0	228	0	932	594	180	41.84
139.6	209.4	0	192	0	1047	806.9	28	23.98
139.6	209.4	0	192	0	1047	806.9	3	8.07

(continued)

(continued)

Cement (kg)	Blast furnace slag (mining) (kg)	Fly ash (kg)	Water (kg)	Water reducer (kg)	Coarse aggregate (kg)	Fine aggregate (kg)	Age (d)	Compressive strength (MPa)
139.6	209.4	0	192	0	1047	806.9	180	44.21
380	0	0	228	0	932	670	365	50.21
380	0	0	228	0	932	670	270	53.3
380	95	0	228	0	932	594	270	41.15
342	38	0	228	0	932	670	180	52.12
427.5	47.5	0	228	0	932	594	28	37.43
475	0	0	228	0	932	594	7	38.6
304	76	0	228	0	932	670	365	55.26
266	114	0	228	0	932	670	365	52.91
198.6	132.4	0	192	0	978.4	825.5	180	41.72
475	0	0	228	0	932	594	270	42.13
190	190	0	228	0	932	670	365	53.69
237.5	237.5	0	228	0	932	594	270	43.24
237.5	237.5	0	228	0	932	594	28	26.37
332.5	142.5	0	228	0	932	594	90	37.72
475	0	0	228	0	932	594	90	42.23
237.5	237.5	0	228	0	932	594	180	36.25
342	38	0	228	0	932	670	90	50.46
427.5	47.5	0	228	0	932	594	365	43.7
237.5	237.5	0	228	0	932	594	365	39
380	0	0	228	0	932	670	180	53.1
427.5	47.5	0	228	0	932	594	90	41.54
427.5	47.5	0	228	0	932	594	7	35.08
349	0	0	192	0	1047	806.9	3	13.20
380	95	0	228	0	932	594	180	40.76
237.5	237.5	0	228	0	932	594	7	26.39
380	95	0	228	0	932	594	7	32.82
332.5	142.5	0	228	0	932	594	180	39.78
190	190	0	228	0	932	670	180	46.93
237.5	237.5	0	228	0	932	594	90	33.12
304	76	0	228	0	932	670	90	49.19
139.6	209.4	0	192	0	1047	806.9	7	20.19
198.6	132.4	0	192	0	978.4	825.5	7	14.64
475	0	0	228	0	932	594	365	41.93

(continued)

(continued)

Cement (kg)	Blast furnace slag (mining) (kg)	Fly ash (kg)	Water (kg)	Water reducer (kg)	Coarse aggregate (kg)	Fine aggregate (kg)	Age (d)	Compressive strength (MPa)
198.6	132.4	0	192	0	978.4	825.5	3	9.13
304	76	0	228	0	932	670	180	62.1
332.5	142.5	0	228	0	932	594	28	33.02
304	76	0	228	0	932	670	270	56.37
266	114	0	228	0	932	670	270	51.73
310	0	0	192	0	971	850.6	3	9.87
190	190	0	228	0	932	670	270	50.66
266	114	0	228	0	932	670	180	44.32
342	38	0	228	0	932	670	270	55.06
139.6	209.4	0	192	0	1047	806.9	360	44.7
332.5	142.5	0	228	0	932	594	7	30.2
190	190	0	228	0	932	670	28	40.86
485	0	0	146	0	1120	800	28	71.99
374	189.2	0	170.1	10.1	926.1	756.7	3	34.4
313.3	262.2	0	175.5	8.6	1046.9	611.8	3	27.38
425	106.3	0	153.5	16.5	852.1	887.1	3	33.4
425	106.3	0	151.4	18.6	936	803.7	3	36.3
375	93.8	0	126.6	23.4	852.1	992.6	3	29
475	118.8	0	181.1	8.9	852.1	781.5	3	37.8
469	117.2	0	137.8	32.2	852.1	840.5	3	40.2
425	106.3	0	153.5	16.5	852.1	887.1	3	33.4
388.6	97.1	0	157.9	12.1	852.1	925.7	3	28.1
531.3	0	0	141.8	28.2	852.1	893.7	3	41.3
425	106.3	0	153.5	16.5	852.1	887.1	3	33.4
318.8	212.5	0	155.7	14.3	852.1	880.4	3	25.2
401.8	94.7	0	147.4	11.4	946.8	852.1	3	41.1
362.6	189	0	164.9	11.6	944.7	755.8	3	35.3
323.7	282.8	0	183.8	10.3	942.7	659.9	3	28.3
379.5	151.2	0	153.9	15.9	1134.3	605	3	28.6
362.6	189	0	164.9	11.6	944.7	755.8	3	35.3
286.3	200.9	0	144.7	11.2	1004.6	803.7	3	24.4
362.6	189	0	164.9	11.6	944.7	755.8	3	35.3
439	177	0	186	11.1	884.9	707.9	3	39.3
389.9	189	0	145.9	22	944.7	755.8	3	40.6

(continued)

(continued)

Cement (kg)	Blast furnace slag (mining) (kg)	Fly ash (kg)	Water (kg)	Water reducer (kg)	Coarse aggregate (kg)	Fine aggregate (kg)	Age (d)	Compressive strength (MPa)
362.6	189	0	164.9	11.6	944.7	755.8	3	30.98
337.9	189	0	174.9	9.5	944.7	755.8	3	24.1
374	189.2	0	170.1	10.1	926.1	756.7	7	46.2
313.3	262.2	0	175.5	8.6	1046.9	611.8	7	42.8
425	106.3	0	153.5	16.5	852.1	887.1	7	49.2
425	106.3	0	151.4	18.6	936	803.7	7	43.29
375	93.8	0	126.6	23.4	852.1	992.6	7	45.7
475	118.8	0	181.1	8.9	852.1	781.5	7	55.6
469	117.2	0	137.8	32.2	852.1	840.5	7	51.2
425	106.3	0	153.5	16.5	852.1	887.1	7	49.2
388.6	97.1	0	157.9	12.1	852.1	925.7	7	34.9
531.3	0	0	141.8	28.2	852.1	893.7	7	46.9
425	106.3	0	153.5	16.5	852.1	887.1	7	49.2
318.8	212.5	0	155.7	14.3	852.1	880.4	7	33.4
401.8	94.7	0	147.4	11.4	946.8	852.1	7	54.1
362.6	189	0	164.9	11.6	944.7	755.8	7	55.9
323.7	282.8	0	183.8	10.3	942.7	659.9	7	49.8
379.5	151.2	0	153.9	15.9	1134.3	605	7	47.1
362.6	189	0	164.9	11.6	944.7	755.8	7	55.9
286.3	200.9	0	144.7	11.2	1004.6	803.7	7	38
362.6	189	0	164.9	11.6	944.7	755.8	7	55.9
439	177	0	186	11.1	884.9	707.9	7	56.1
389.9	189	0	145.9	22	944.7	755.8	7	59.09
362.6	189	0	164.9	11.6	944.7	755.8	7	22.9
337.9	189	0	174.9	9.5	944.7	755.8	7	35.1
374	189.2	0	170.1	10.1	926.1	756.7	28	61.09
313.3	262.2	0	175.5	8.6	1046.9	611.8	28	59.8
425	106.3	0	153.5	16.5	852.1	887.1	28	60.29
425	106.3	0	151.4	18.6	936	803.7	28	61.8
375	93.8	0	126.6	23.4	852.1	992.6	28	56.7
475	118.8	0	181.1	8.9	852.1	781.5	28	68.3
469	117.2	0	137.8	32.2	852.1	840.5	28	66.9
425	106.3	0	153.5	16.5	852.1	887.1	28	60.29
388.6	97.1	0	157.9	12.1	852.1	925.7	28	50.7

(continued)

(continued)

Cement (kg)	Blast furnace slag (mining) (kg)	Fly ash (kg)	Water (kg)	Water reducer (kg)	Coarse aggregate (kg)	Fine aggregate (kg)	Age (d)	Compressive strength (MPa)
531.3	0	0	141.8	28.2	852.1	893.7	28	56.4
425	106.3	0	153.5	16.5	852.1	887.1	28	60.29
318.8	212.5	0	155.7	14.3	852.1	880.4	28	50.21
401.8	94.7	0	147.4	11.4	946.8	852.1	28	68.5
362.6	189	0	164.9	11.6	944.7	755.8	28	71.3
323.7	282.8	0	183.8	10.3	942.7	659.9	28	74.7
379.5	151.2	0	153.9	15.9	1134.3	605	28	52.2
362.6	189	0	164.9	11.6	944.7	755.8	28	71.3
286.3	200.9	0	144.7	11.2	1004.6	803.7	28	67.7
362.6	189	0	164.9	11.6	944.7	755.8	28	71.3
439	177	0	186	11.1	884.9	707.9	28	66
389.9	189	0	145.9	22	944.7	755.8	28	74.5
362.6	189	0	164.9	11.6	944.7	755.8	28	71.3
337.9	189	0	174.9	9.5	944.7	755.8	28	49.9
374	189.2	0	170.1	10.1	926.1	756.7	56	63.4
313.3	262.2	0	175.5	8.6	1046.9	611.8	56	64.9
425	106.3	0	153.5	16.5	852.1	887.1	56	64.3
425	106.3	0	151.4	18.6	936	803.7	56	64.9
375	93.8	0	126.6	23.4	852.1	992.6	56	60.2
475	118.8	0	181.1	8.9	852.1	781.5	56	72.3
469	117.2	0	137.8	32.2	852.1	840.5	56	69.3
425	106.3	0	153.5	16.5	852.1	887.1	56	64.3
388.6	97.1	0	157.9	12.1	852.1	925.7	56	55.2
531.3	0	0	141.8	28.2	852.1	893.7	56	58.8
425	106.3	0	153.5	16.5	852.1	887.1	56	64.3
318.8	212.5	0	155.7	14.3	852.1	880.4	56	66.1
401.8	94.7	0	147.4	11.4	946.8	852.1	56	73.7
362.6	189	0	164.9	11.6	944.7	755.8	56	77.3
323.7	282.8	0	183.8	10.3	942.7	659.9	56	80.2
379.5	151.2	0	153.9	15.9	1134.3	605	56	54.9
362.6	189	0	164.9	11.6	944.7	755.8	56	77.3
286.3	200.9	0	144.7	11.2	1004.6	803.7	56	72.99
362.6	189	0	164.9	11.6	944.7	755.8	56	77.3
439	177	0	186	11.1	884.9	707.9	56	71.7

(continued)

(continued)

Cement (kg)	Blast furnace slag (mining) (kg)	Fly ash (kg)	Water (kg)	Water reducer (kg)	Coarse aggregate (kg)	Fine aggregate (kg)	Age (d)	Compressive strength (MPa)
389.9	189	0	145.9	22	944.7	755.8	56	79.4
362.6	189	0	164.9	11.6	944.7	755.8	56	77.3
337.9	189	0	174.9	9.5	944.7	755.8	56	59.89
374	189.2	0	170.1	10.1	926.1	756.7	91	64.9
313.3	262.2	0	175.5	8.6	1046.9	611.8	91	66.6
425	106.3	0	153.5	16.5	852.1	887.1	91	65.2
425	106.3	0	151.4	18.6	936	803.7	91	66.7
375	93.8	0	126.6	23.4	852.1	992.6	91	62.5
475	118.8	0	181.1	8.9	852.1	781.5	91	74.19
469	117.2	0	137.8	32.2	852.1	840.5	91	70.7
425	106.3	0	153.5	16.5	852.1	887.1	91	65.2
388.6	97.1	0	157.9	12.1	852.1	925.7	91	57.6
531.3	0	0	141.8	28.2	852.1	893.7	91	59.2
425	106.3	0	153.5	16.5	852.1	887.1	91	65.2
318.8	212.5	0	155.7	14.3	852.1	880.4	91	68.1
401.8	94.7	0	147.4	11.4	946.8	852.1	91	75.5
362.6	189	0	164.9	11.6	944.7	755.8	91	79.3
379.5	151.2	0	153.9	15.9	1134.3	605	91	56.5
362.6	189	0	164.9	11.6	944.7	755.8	91	79.3
286.3	200.9	0	144.7	11.2	1004.6	803.7	91	76.8
362.6	189	0	164.9	11.6	944.7	755.8	91	79.3
439	177	0	186	11.1	884.9	707.9	91	73.3
389.9	189	0	145.9	22	944.7	755.8	91	82.6
362.6	189	0	164.9	11.6	944.7	755.8	91	79.3
337.9	189	0	174.9	9.5	944.7	755.8	91	67.8
222.4	0	96.7	189.3	4.5	967.1	870.3	3	11.58
222.4	0	96.7	189.3	4.5	967.1	870.3	14	24.45
222.4	0	96.7	189.3	4.5	967.1	870.3	28	24.89
222.4	0	96.7	189.3	4.5	967.1	870.3	56	29.45
222.4	0	96.7	189.3	4.5	967.1	870.3	100	40.71
233.8	0	94.6	197.9	4.6	947	852.2	3	10.38
233.8	0	94.6	197.9	4.6	947	852.2	14	22.14
233.8	0	94.6	197.9	4.6	947	852.2	28	22.84
233.8	0	94.6	197.9	4.6	947	852.2	56	27.66

(continued)

(continued)

Cement (kg)	Blast furnace slag (mining) (kg)	Fly ash (kg)	Water (kg)	Water reducer (kg)	Coarse aggregate (kg)	Fine aggregate (kg)	Age (d)	Compressive strength (MPa)
233.8	0	94.6	197.9	4.6	947	852.2	100	34.56
194.7	0	100.5	165.6	7.5	1006.4	905.9	3	17.1
194.7	0	100.5	165.6	7.5	1006.4	905.9	14	24.99
194.7	0	100.5	165.6	7.5	1006.4	905.9	28	25.72
194.7	0	100.5	165.6	7.5	1006.4	905.9	56	33.96
194.7	0	100.5	165.6	7.5	1006.4	905.9	100	37.34
190.7	0	125.4	162.1	7.8	1090	804	3	15.04
190.7	0	125.4	162.1	7.8	1090	804	14	21.06
190.7	0	125.4	162.1	7.8	1090	804	28	26.4
190.7	0	125.4	162.1	7.8	1090	804	56	30.2
190.7	0	125.4	162.1	7.8	1090	804	100	40.57
212.1	0	121.6	180.3	5.7	1057.6	779.3	3	12.47
212.1	0	121.6	180.3	5.7	1057.6	779.3	14	20.92
212.1	0	121.6	180.3	5.7	1057.6	779.3	28	24.9
212.1	0	121.6	180.3	5.7	1057.6	779.3	56	34.2
212.1	0	121.6	180.3	5.7	1057.6	779.3	100	39.61
230	0	118.3	195.5	4.6	1029.4	758.6	3	10.03
230	0	118.3	195.5	4.6	1029.4	758.6	14	20.19
230	0	118.3	195.5	4.6	1029.4	758.6	28	24.48
230	0	118.3	195.5	4.6	1029.4	758.6	56	31.54
230	0	118.3	195.5	4.6	1029.4	758.6	100	35.34
190.3	0	125.2	161.9	9.9	1088.1	802.6	3	9.45
190.3	0	125.2	161.9	9.9	1088.1	802.6	14	22.72
190.3	0	125.2	161.9	9.9	1088.1	802.6	28	28.47
190.3	0	125.2	161.9	9.9	1088.1	802.6	56	38.56
190.3	0	125.2	161.9	9.9	1088.1	802.6	100	44.32
166.1	0	163.3	176.5	4.5	1058.6	780.1	3	10.76
166.1	0	163.3	176.5	4.5	1058.6	780.1	14	25.48
166.1	0	163.3	176.5	4.5	1058.6	780.1	28	21.54
166.1	0	163.3	176.5	4.5	1058.6	780.1	56	28.63
166.1	0	163.3	176.5	4.5	1058.6	780.1	100	33.54
168	42.1	163.8	121.8	5.7	1058.7	780.1	3	7.75
168	42.1	163.8	121.8	5.7	1058.7	780.1	14	17.82
168	42.1	163.8	121.8	5.7	1058.7	780.1	28	26.39

(continued)

(continued)

Cement (kg)	Blast furnace slag (mining) (kg)	Fly ash (kg)	Water (kg)	Water reducer (kg)	Coarse aggregate (kg)	Fine aggregate (kg)	Age (d)	Compressive strength (MPa)
168	42.1	163.8	121.8	5.7	1058.7	780.1	56	32.85
168	42.1	163.8	121.8	5.7	1058.7	780.1	100	39.23
213.7	98.1	24.5	181.7	6.9	1065.8	785.4	3	18
213.7	98.1	24.5	181.7	6.9	1065.8	785.4	14	30.39
213.7	98.1	24.5	181.7	6.9	1065.8	785.4	28	45.71
213.7	98.1	24.5	181.7	6.9	1065.8	785.4	56	50.77
213.7	98.1	24.5	181.7	6.9	1065.8	785.4	100	53.9
213.8	98.1	24.5	181.7	6.7	1066	785.5	3	13.18
213.8	98.1	24.5	181.7	6.7	1066	785.5	14	17.84
213.8	98.1	24.5	181.7	6.7	1066	785.5	28	40.23
213.8	98.1	24.5	181.7	6.7	1066	785.5	56	47.13
213.8	98.1	24.5	181.7	6.7	1066	785.5	100	49.97
229.7	0	118.2	195.2	6.1	1028.1	757.6	3	13.36
229.7	0	118.2	195.2	6.1	1028.1	757.6	14	22.32
229.7	0	118.2	195.2	6.1	1028.1	757.6	28	24.54
229.7	0	118.2	195.2	6.1	1028.1	757.6	56	31.35
229.7	0	118.2	195.2	6.1	1028.1	757.6	100	40.86
238.1	0	94.1	186.7	7	949.9	847	3	19.93
238.1	0	94.1	186.7	7	949.9	847	14	25.69
238.1	0	94.1	186.7	7	949.9	847	28	30.23
238.1	0	94.1	186.7	7	949.9	847	56	39.59
238.1	0	94.1	186.7	7	949.9	847	100	44.3
250	0	95.7	187.4	5.5	956.9	861.2	3	13.82
250	0	95.7	187.4	5.5	956.9	861.2	14	24.92
250	0	95.7	187.4	5.5	956.9	861.2	28	29.22
250	0	95.7	187.4	5.5	956.9	861.2	56	38.33
250	0	95.7	187.4	5.5	956.9	861.2	100	42.35
212.5	0	100.4	159.3	8.7	1007.8	903.6	3	13.54
212.5	0	100.4	159.3	8.7	1007.8	903.6	14	26.31
212.5	0	100.4	159.3	8.7	1007.8	903.6	28	31.64
212.5	0	100.4	159.3	8.7	1007.8	903.6	56	42.55
212.5	0	100.4	159.3	8.7	1007.8	903.6	100	42.92
212.6	0	100.4	159.4	10.4	1003.8	903.8	3	13.33
212.6	0	100.4	159.4	10.4	1003.8	903.8	14	25.37

(continued)

(continued)

Cement (kg)	Blast furnace slag (mining) (kg)	Fly ash (kg)	Water (kg)	Water reducer (kg)	Coarse aggregate (kg)	Fine aggregate (kg)	Age (d)	Compressive strength (MPa)
212.6	0	100.4	159.4	10.4	1003.8	903.8	28	37.4
212.6	0	100.4	159.4	10.4	1003.8	903.8	56	44.4
212.6	0	100.4	159.4	10.4	1003.8	903.8	100	47.74
212	0	124.8	159	7.8	1085.4	799.5	3	19.52
212	0	124.8	159	7.8	1085.4	799.5	14	31.35
212	0	124.8	159	7.8	1085.4	799.5	28	38.5
212	0	124.8	159	7.8	1085.4	799.5	56	45.08
212	0	124.8	159	7.8	1085.4	799.5	100	47.82
231.8	0	121.6	174	6.7	1056.4	778.5	3	15.44
231.8	0	121.6	174	6.7	1056.4	778.5	14	26.77
231.8	0	121.6	174	6.7	1056.4	778.5	28	33.73
231.8	0	121.6	174	6.7	1056.4	778.5	56	42.7
231.8	0	121.6	174	6.7	1056.4	778.5	100	45.84
251.4	0	118.3	188.5	5.8	1028.4	757.7	3	17.22
251.4	0	118.3	188.5	5.8	1028.4	757.7	14	29.93
251.4	0	118.3	188.5	5.8	1028.4	757.7	28	29.65
251.4	0	118.3	188.5	5.8	1028.4	757.7	56	36.97
251.4	0	118.3	188.5	5.8	1028.4	757.7	100	43.58
251.4	0	118.3	188.5	6.4	1028.4	757.7	3	13.12
251.4	0	118.3	188.5	6.4	1028.4	757.7	14	24.43
251.4	0	118.3	188.5	6.4	1028.4	757.7	28	32.66
251.4	0	118.3	188.5	6.4	1028.4	757.7	56	36.64
251.4	0	118.3	188.5	6.4	1028.4	757.7	100	44.21
181.4	0	167	169.6	7.6	1055.6	777.8	3	13.62
181.4	0	167	169.6	7.6	1055.6	777.8	14	21.6
181.4	0	167	169.6	7.6	1055.6	777.8	28	27.77
181.4	0	167	169.6	7.6	1055.6	777.8	56	35.57
181.4	0	167	169.6	7.6	1055.6	777.8	100	45.37
182	45.2	122	170.2	8.2	1059.4	780.7	3	7.32
182	45.2	122	170.2	8.2	1059.4	780.7	14	21.5
182	45.2	122	170.2	8.2	1059.4	780.7	28	31.27
182	45.2	122	170.2	8.2	1059.4	780.7	56	43.5
182	45.2	122	170.2	8.2	1059.4	780.7	100	48.67
168.9	42.2	124.3	158.3	10.8	1080.8	796.2	3	7.4

(continued)

(continued)

Cement (kg)	Blast furnace slag (mining) (kg)	Fly ash (kg)	Water (kg)	Water reducer (kg)	Coarse aggregate (kg)	Fine aggregate (kg)	Age (d)	Compressive strength (MPa)
168.9	42.2	124.3	158.3	10.8	1080.8	796.2	14	23.51
168.9	42.2	124.3	158.3	10.8	1080.8	796.2	28	31.12
168.9	42.2	124.3	158.3	10.8	1080.8	796.2	56	39.15
168.9	42.2	124.3	158.3	10.8	1080.8	796.2	100	48.15
290.4	0	96.2	168.1	9.4	961.2	865	3	22.5
290.4	0	96.2	168.1	9.4	961.2	865	14	34.67
290.4	0	96.2	168.1	9.4	961.2	865	28	34.74
290.4	0	96.2	168.1	9.4	961.2	865	56	45.08
290.4	0	96.2	168.1	9.4	961.2	865	100	48.97
277.1	0	97.4	160.6	11.8	973.9	875.6	3	23.14
277.1	0	97.4	160.6	11.8	973.9	875.6	14	41.89
277.1	0	97.4	160.6	11.8	973.9	875.6	28	48.28
277.1	0	97.4	160.6	11.8	973.9	875.6	56	51.04
277.1	0	97.4	160.6	11.8	973.9	875.6	100	55.64
295.7	0	95.6	171.5	8.9	955.1	859.2	3	22.95
295.7	0	95.6	171.5	8.9	955.1	859.2	14	35.23
295.7	0	95.6	171.5	8.9	955.1	859.2	28	39.94
295.7	0	95.6	171.5	8.9	955.1	859.2	56	48.72
295.7	0	95.6	171.5	8.9	955.1	859.2	100	52.04
251.8	0	99.9	146.1	12.4	1006	899.8	3	21.02
251.8	0	99.9	146.1	12.4	1006	899.8	14	33.36
251.8	0	99.9	146.1	12.4	1006	899.8	28	33.94
251.8	0	99.9	146.1	12.4	1006	899.8	56	44.14
251.8	0	99.9	146.1	12.4	1006	899.8	100	45.37
249.1	0	98.8	158.1	12.8	987.8	889	3	15.36
249.1	0	98.8	158.1	12.8	987.8	889	14	28.68
249.1	0	98.8	158.1	12.8	987.8	889	28	30.85
249.1	0	98.8	158.1	12.8	987.8	889	56	42.03
249.1	0	98.8	158.1	12.8	987.8	889	100	51.06
252.3	0	98.8	146.3	14.2	987.8	889	3	21.78
252.3	0	98.8	146.3	14.2	987.8	889	14	42.29
252.3	0	98.8	146.3	14.2	987.8	889	28	50.6
252.3	0	98.8	146.3	14.2	987.8	889	56	55.83
252.3	0	98.8	146.3	14.2	987.8	889	100	60.95

(continued)

Appendix: Summary of Hardened Concrete Compressive Test Data ...

(continued)

Cement (kg)	Blast furnace slag (mining) (kg)	Fly ash (kg)	Water (kg)	Water reducer (kg)	Coarse aggregate (kg)	Fine aggregate (kg)	Age (d)	Compressive strength (MPa)
246.8	0	125.1	143.3	12	1086.8	800.9	3	23.52
246.8	0	125.1	143.3	12	1086.8	800.9	14	42.22
246.8	0	125.1	143.3	12	1086.8	800.9	28	52.5
246.8	0	125.1	143.3	12	1086.8	800.9	56	60.32
246.8	0	125.1	143.3	12	1086.8	800.9	100	66.42
275.1	0	121.4	159.5	9.9	1053.6	777.5	3	23.8
275.1	0	121.4	159.5	9.9	1053.6	777.5	14	38.77
275.1	0	121.4	159.5	9.9	1053.6	777.5	28	51.33
275.1	0	121.4	159.5	9.9	1053.6	777.5	56	56.85
275.1	0	121.4	159.5	9.9	1053.6	777.5	100	58.61
297.2	0	117.5	174.8	9.5	1022.8	753.5	3	21.91
297.2	0	117.5	174.8	9.5	1022.8	753.5	14	36.99
297.2	0	117.5	174.8	9.5	1022.8	753.5	28	47.4
297.2	0	117.5	174.8	9.5	1022.8	753.5	56	51.96
297.2	0	117.5	174.8	9.5	1022.8	753.5	100	56.74
213.7	0	174.7	154.8	10.2	1053.5	776.4	3	17.57
213.7	0	174.7	154.8	10.2	1053.5	776.4	14	33.73
213.7	0	174.7	154.8	10.2	1053.5	776.4	28	40.15
213.7	0	174.7	154.8	10.2	1053.5	776.4	56	46.64
213.7	0	174.7	154.8	10.2	1053.5	776.4	100	50.08
213.5	0	174.2	154.6	11.7	1052.3	775.5	3	17.37
213.5	0	174.2	154.6	11.7	1052.3	775.5	14	33.7
213.5	0	174.2	154.6	11.7	1052.3	775.5	28	45.94
213.5	0	174.2	154.6	11.7	1052.3	775.5	56	51.43
213.5	0	174.2	154.6	11.7	1052.3	775.5	100	59.3
277.2	97.8	24.5	160.7	11.2	1061.7	782.5	3	30.45
277.2	97.8	24.5	160.7	11.2	1061.7	782.5	14	47.71
277.2	97.8	24.5	160.7	11.2	1061.7	782.5	28	63.14
277.2	97.8	24.5	160.7	11.2	1061.7	782.5	56	66.82
277.2	97.8	24.5	160.7	11.2	1061.7	782.5	100	66.95
218.2	54.6	123.8	140.8	11.9	1075.7	792.7	3	27.42
218.2	54.6	123.8	140.8	11.9	1075.7	792.7	14	35.96
218.2	54.6	123.8	140.8	11.9	1075.7	792.7	28	55.51
218.2	54.6	123.8	140.8	11.9	1075.7	792.7	56	61.99

(continued)

(continued)

Cement (kg)	Blast furnace slag (mining) (kg)	Fly ash (kg)	Water (kg)	Water reducer (kg)	Coarse aggregate (kg)	Fine aggregate (kg)	Age (d)	Compressive strength (MPa)
218.2	54.6	123.8	140.8	11.9	1075.7	792.7	100	63.53
214.9	53.8	121.9	155.6	9.6	1014.3	780.6	3	18.02
214.9	53.8	121.9	155.6	9.6	1014.3	780.6	14	38.6
214.9	53.8	121.9	155.6	9.6	1014.3	780.6	28	52.2
214.9	53.8	121.9	155.6	9.6	1014.3	780.6	56	53.96
214.9	53.8	121.9	155.6	9.6	1014.3	780.6	100	56.63
218.9	0	124.1	158.5	11.3	1078.7	794.9	3	15.34
218.9	0	124.1	158.5	11.3	1078.7	794.9	14	26.05
218.9	0	124.1	158.5	11.3	1078.7	794.9	28	30.22
218.9	0	124.1	158.5	11.3	1078.7	794.9	56	37.27
218.9	0	124.1	158.5	11.3	1078.7	794.9	100	46.23
376	0	0	214.6	0	1003.5	762.4	3	16.28
376	0	0	214.6	0	1003.5	762.4	14	25.62
376	0	0	214.6	0	1003.5	762.4	28	31.97
376	0	0	214.6	0	1003.5	762.4	56	36.3
376	0	0	214.6	0	1003.5	762.4	100	43.06
500	0	0	140	4	966	853	28	67.57
475	0	59	142	1.9	1098	641	28	57.23
315	137	0	145	5.9	1130	745	28	81.75
505	0	60	195	0	1030	630	28	64.02
451	0	0	165	11.3	1030	745	28	78.8
516	0	0	162	8.2	801	802	28	41.37
520	0	0	170	5.2	855	855	28	60.28
528	0	0	185	6.9	920	720	28	56.83
520	0	0	175	5.2	870	805	28	51.02
385	0	136	158	20	903	768	28	55.55
500.1	0	0	200	3	1124.4	613.2	28	44.13
450.1	50	0	200	3	1124.4	613.2	28	39.38
397	17.2	158	167	20.8	967	633	28	55.65
333	17.5	163	167	17.9	996	652	28	47.28
334	17.6	158	189	15.3	967	633	28	44.33
405	0	0	175	0	1120	695	28	52.3
200	200	0	190	0	1145	660	28	49.25
516	0	0	162	8.3	801	802	28	41.37

(continued)

(continued)

Cement (kg)	Blast furnace slag (mining) (kg)	Fly ash (kg)	Water (kg)	Water reducer (kg)	Coarse aggregate (kg)	Fine aggregate (kg)	Age (d)	Compressive strength (MPa)
145	116	119	184	5.7	833	880	28	29.16
160	128	122	182	6.4	824	879	28	39.4
234	156	0	189	5.9	981	760	28	39.3
250	180	95	159	9.5	860	800	28	67.87
475	0	0	162	9.5	1044	662	28	58.52
285	190	0	163	7.6	1031	685	28	53.58
356	119	0	160	9	1061	657	28	59
275	180	120	162	10.4	830	765	28	76.24
500	0	0	151	9	1033	655	28	69.84
165	0	143.6	163.8	0	1005.6	900.9	3	14.4
165	128.5	132.1	175.1	8.1	1005.8	746.6	3	19.42
178	129.8	118.6	179.9	3.6	1007.3	746.8	3	20.73
167.4	129.9	128.6	175.5	7.8	1006.3	746.6	3	14.94
172.4	13.6	172.4	156.8	4.1	1006.3	856.4	3	21.29
173.5	50.1	173.5	164.8	6.5	1006.2	793.5	3	23.08
167	75.4	167	164	7.9	1007.3	770.1	3	15.52
173.8	93.4	159.9	172.3	9.7	1007.2	746.6	3	15.82
190.3	0	125.2	166.6	9.9	1079	798.9	3	12.55
250	0	95.7	191.8	5.3	948.9	857.2	3	8.49
213.5	0	174.2	159.2	11.7	1043.6	771.9	3	15.61
194.7	0	100.5	170.2	7.5	998	901.8	3	12.18
251.4	0	118.3	192.9	5.8	1043.6	754.3	3	11.98
165	0	143.6	163.8	0	1005.6	900.9	14	16.88
165	128.5	132.1	175.1	8.1	1005.8	746.6	14	33.09
178	129.8	118.6	179.9	3.6	1007.3	746.8	14	34.24
167.4	129.9	128.6	175.5	7.8	1006.3	746.6	14	31.81
172.4	13.6	172.4	156.8	4.1	1006.3	856.4	14	29.75
173.5	50.1	173.5	164.8	6.5	1006.2	793.5	14	33.01
167	75.4	167	164	7.9	1007.3	770.1	14	32.9
173.8	93.4	159.9	172.3	9.7	1007.2	746.6	14	29.55
190.3	0	125.2	166.6	9.9	1079	798.9	14	19.42
250	0	95.7	191.8	5.3	948.9	857.2	14	24.66
213.5	0	174.2	159.2	11.7	1043.6	771.9	14	29.59
194.7	0	100.5	170.2	7.5	998	901.8	14	24.28

(continued)

(continued)

Cement (kg)	Blast furnace slag (mining) (kg)	Fly ash (kg)	Water (kg)	Water reducer (kg)	Coarse aggregate (kg)	Fine aggregate (kg)	Age (d)	Compressive strength (MPa)
251.4	0	118.3	192.9	5.8	1043.6	754.3	14	20.73
165	0	143.6	163.8	0	1005.6	900.9	28	26.2
165	128.5	132.1	175.1	8.1	1005.8	746.6	28	46.39
178	129.8	118.6	179.9	3.6	1007.3	746.8	28	39.16
167.4	129.9	128.6	175.5	7.8	1006.3	746.6	28	41.2
172.4	13.6	172.4	156.8	4.1	1006.3	856.4	28	33.69
173.5	50.1	173.5	164.8	6.5	1006.2	793.5	28	38.2
167	75.4	167	164	7.9	1007.3	770.1	28	41.41
173.8	93.4	159.9	172.3	9.7	1007.2	746.6	28	37.81
190.3	0	125.2	166.6	9.9	1079	798.9	28	24.85
250	0	95.7	191.8	5.3	948.9	857.2	28	27.22
213.5	0	174.2	159.2	11.7	1043.6	771.9	28	44.64
194.7	0	100.5	170.2	7.5	998	901.8	28	37.27
251.4	0	118.3	192.9	5.8	1043.6	754.3	28	33.27
165	0	143.6	163.8	0	1005.6	900.9	56	36.56
165	128.5	132.1	175.1	8.1	1005.8	746.6	56	53.72
178	129.8	118.6	179.9	3.6	1007.3	746.8	56	48.59
167.4	129.9	128.6	175.5	7.8	1006.3	746.6	56	51.72
172.4	13.6	172.4	156.8	4.1	1006.3	856.4	56	35.85
173.5	50.1	173.5	164.8	6.5	1006.2	793.5	56	53.77
167	75.4	167	164	7.9	1007.3	770.1	56	53.46
173.8	93.4	159.9	172.3	9.7	1007.2	746.6	56	48.99
190.3	0	125.2	166.6	9.9	1079	798.9	56	31.72
250	0	95.7	191.8	5.3	948.9	857.2	56	39.64
213.5	0	174.2	159.2	11.7	1043.6	771.9	56	51.26
194.7	0	100.5	170.2	7.5	998	901.8	56	43.39
251.4	0	118.3	192.9	5.8	1043.6	754.3	56	39.27
165	0	143.6	163.8	0	1005.6	900.9	100	37.96
165	128.5	132.1	175.1	8.1	1005.8	746.6	100	55.02
178	129.8	118.6	179.9	3.6	1007.3	746.8	100	49.99
167.4	129.9	128.6	175.5	7.8	1006.3	746.6	100	53.66
172.4	13.6	172.4	156.8	4.1	1006.3	856.4	100	37.68
173.5	50.1	173.5	164.8	6.5	1006.2	793.5	100	56.06
167	75.4	167	164	7.9	1007.3	770.1	100	56.81

(continued)

(continued)

Cement (kg)	Blast furnace slag (mining) (kg)	Fly ash (kg)	Water (kg)	Water reducer (kg)	Coarse aggregate (kg)	Fine aggregate (kg)	Age (d)	Compressive strength (MPa)
173.8	93.4	159.9	172.3	9.7	1007.2	746.6	100	50.94
190.3	0	125.2	166.6	9.9	1079	798.9	100	33.56
250	0	95.7	191.8	5.3	948.9	857.2	100	41.16
213.5	0	174.2	159.2	11.7	1043.6	771.9	100	52.96
194.7	0	100.5	170.2	7.5	998	901.8	100	44.28
251.4	0	118.3	192.9	5.8	1043.6	754.3	100	40.15
446	24	79	162	11.6	967	712	28	57.03
446	24	79	162	11.6	967	712	28	44.42
446	24	79	162	11.6	967	712	28	51.02
446	24	79	162	10.3	967	712	28	53.39
446	24	79	162	11.6	967	712	3	35.36
446	24	79	162	11.6	967	712	3	25.02
446	24	79	162	11.6	967	712	3	23.35
446	24	79	162	11.6	967	712	7	52.01
446	24	79	162	11.6	967	712	7	38.02
446	24	79	162	11.6	967	712	7	39.3
446	24	79	162	11.6	967	712	56	61.07
446	24	79	162	11.6	967	712	56	56.14
446	24	79	162	11.6	967	712	56	55.25
446	24	79	162	10.3	967	712	56	54.77
387	20	94	157	14.3	938	845	28	50.24
387	20	94	157	13.9	938	845	28	46.68
387	20	94	157	11.6	938	845	28	46.68
387	20	94	157	14.3	938	845	3	22.75
387	20	94	157	13.9	938	845	3	25.51
387	20	94	157	11.6	938	845	3	34.77
387	20	94	157	14.3	938	845	7	36.84
387	20	94	157	13.9	938	845	7	45.9
387	20	94	157	11.6	938	845	7	41.67
387	20	94	157	14.3	938	845	56	56.34
387	20	94	157	13.9	938	845	56	47.97
387	20	94	157	11.6	938	845	56	61.46
355	19	97	145	13.1	967	871	28	44.03
355	19	97	145	12.3	967	871	28	55.45

(continued)

(continued)

Cement (kg)	Blast furnace slag (mining) (kg)	Fly ash (kg)	Water (kg)	Water reducer (kg)	Coarse aggregate (kg)	Fine aggregate (kg)	Age (d)	Compressive strength (MPa)
491	26	123	210	3.9	882	699	28	55.55
491	26	123	201	3.9	822	699	28	57.92
491	26	123	210	3.9	882	699	3	25.61
491	26	123	210	3.9	882	699	7	33.49
491	26	123	210	3.9	882	699	56	59.59
491	26	123	201	3.9	822	699	3	29.55
491	26	123	201	3.9	822	699	7	37.92
491	26	123	201	3.9	822	699	56	61.86
424	22	132	178	8.5	822	750	28	62.05
424	22	132	178	8.5	882	750	3	32.01
424	22	132	168	8.9	822	750	28	72.1
424	22	132	178	8.5	822	750	7	39
424	22	132	178	8.5	822	750	56	65.7
424	22	132	168	8.9	822	750	3	32.11
424	22	132	168	8.9	822	750	7	40.29
424	22	132	168	8.9	822	750	56	74.36
202	11	141	206	1.7	942	801	28	21.97
202	11	141	206	1.7	942	801	3	9.85
202	11	141	206	1.7	942	801	7	15.07
202	11	141	206	1.7	942	801	56	23.25
284	15	141	179	5.5	842	801	28	43.73
284	15	141	179	5.5	842	801	3	13.4
284	15	141	179	5.5	842	801	7	24.13
284	15	141	179	5.5	842	801	56	44.52
359	19	141	154	10.9	942	801	28	62.94
359	19	141	154	10.9	942	801	28	59.49
359	19	141	154	10.9	942	801	3	25.12
359	19	141	154	10.9	942	801	3	23.64
359	19	141	154	10.9	942	801	7	35.75
359	19	141	154	10.9	942	801	7	38.61
359	19	141	154	10.9	942	801	56	68.75
359	19	141	154	10.9	942	801	56	66.78
436	0	0	218	0	838.4	719.7	28	23.85
289	0	0	192	0	913.2	895.3	90	32.07

(continued)

(continued)

Cement (kg)	Blast furnace slag (mining) (kg)	Fly ash (kg)	Water (kg)	Water reducer (kg)	Coarse aggregate (kg)	Fine aggregate (kg)	Age (d)	Compressive strength (MPa)
289	0	0	192	0	913.2	895.3	3	11.65
393	0	0	192	0	940.6	785.6	3	19.2
393	0	0	192	0	940.6	785.6	90	48.85
393	0	0	192	0	940.6	785.6	28	39.6
480	0	0	192	0	936.2	712.2	28	43.94
480	0	0	192	0	936.2	712.2	7	34.57
480	0	0	192	0	936.2	712.2	90	54.32
480	0	0	192	0	936.2	712.2	3	24.4
333	0	0	192	0	931.2	842.6	3	15.62
255	0	0	192	0	889.8	945	90	21.86
255	0	0	192	0	889.8	945	7	10.22
289	0	0	192	0	913.2	895.3	7	14.6
255	0	0	192	0	889.8	945	28	18.75
333	0	0	192	0	931.2	842.6	28	31.97
333	0	0	192	0	931.2	842.6	7	23.4
289	0	0	192	0	913.2	895.3	28	25.57
333	0	0	192	0	931.2	842.6	90	41.68
393	0	0	192	0	940.6	785.6	7	27.74
255	0	0	192	0	889.8	945	3	8.2
158.8	238.2	0	185.7	0	1040.6	734.3	7	9.62
239.6	359.4	0	185.7	0	941.6	664.3	7	25.42
238.2	158.8	0	185.7	0	1040.6	734.3	7	15.69
181.9	272.8	0	185.7	0	1012.4	714.3	28	27.94
193.5	290.2	0	185.7	0	998.2	704.3	28	32.63
255.5	170.3	0	185.7	0	1026.6	724.3	7	17.24
272.8	181.9	0	185.7	0	1012.4	714.3	7	19.77
239.6	359.4	0	185.7	0	941.6	664.3	28	39.44
220.8	147.2	0	185.7	0	1055	744.3	28	25.75
397	0	0	185.7	0	1040.6	734.3	28	33.08
382.5	0	0	185.7	0	1047.8	739.3	28	24.07
210.7	316.1	0	185.7	0	977	689.3	7	21.82
158.8	238.2	0	185.7	0	1040.6	734.3	28	21.07
295.8	0	0	185.7	0	1091.4	769.3	7	14.84
255.5	170.3	0	185.7	0	1026.6	724.3	28	32.05

(continued)

(continued)

Cement (kg)	Blast furnace slag (mining) (kg)	Fly ash (kg)	Water (kg)	Water reducer (kg)	Coarse aggregate (kg)	Fine aggregate (kg)	Age (d)	Compressive strength (MPa)
203.5	135.7	0	185.7	0	1076.2	759.3	7	11.96
397	0	0	185.7	0	1040.6	734.3	7	25.45
381.4	0	0	185.7	0	1104.6	784.3	28	22.49
295.8	0	0	185.7	0	1091.4	769.3	28	25.22
228	342.1	0	185.7	0	955.8	674.3	28	39.7
220.8	147.2	0	185.7	0	1055	744.3	7	13.09
316.1	210.7	0	185.7	0	977	689.3	28	38.7
135.7	203.5	0	185.7	0	1076.2	759.3	7	7.51
238.1	0	0	185.7	0	1118.8	789.3	28	17.58
339.2	0	0	185.7	0	1069.2	754.3	7	21.18
135.7	203.5	0	185.7	0	1076.2	759.3	28	18.2
193.5	290.2	0	185.7	0	998.2	704.3	7	17.2
203.5	135.7	0	185.7	0	1076.2	759.3	28	22.63
290.2	193.5	0	185.7	0	998.2	704.3	7	21.86
181.9	272.8	0	185.7	0	1012.4	714.3	7	12.37
170.3	155.5	0	185.7	0	1026.6	724.3	28	25.73
210.7	316.1	0	185.7	0	977	689.3	28	37.81
228	342.1	0	185.7	0	955.8	674.3	7	21.92
290.2	193.5	0	185.7	0	998.2	704.3	28	33.04
381.4	0	0	185.7	0	1104.6	784.3	7	14.54
238.2	158.8	0	185.7	0	1040.6	734.3	28	26.91
186.2	124.1	0	185.7	0	1083.4	764.3	7	8
339.2	0	0	185.7	0	1069.2	754.3	28	31.9
238.1	0	0	185.7	0	1118.8	789.3	7	10.34
252.5	0	0	185.7	0	1111.6	784.3	28	19.77
382.5	0	0	185.7	0	1047.8	739.3	28	37.44
252.5	0	0	185.7	0	1111.6	784.3	7	11.48
316.1	210.7	0	185.7	0	977	689.3	7	24.44
186.2	124.1	0	185.7	0	1083.4	764.3	28	17.6
170.3	155.5	0	185.7	0	1026.6	724.3	7	10.73
272.8	181.9	0	185.7	0	1012.4	714.3	28	31.38
339	0	0	197	0	968	781	3	13.22
339	0	0	197	0	968	781	7	20.97
339	0	0	197	0	968	781	14	27.04

(continued)

(continued)

Cement (kg)	Blast furnace slag (mining) (kg)	Fly ash (kg)	Water (kg)	Water reducer (kg)	Coarse aggregate (kg)	Fine aggregate (kg)	Age (d)	Compressive strength (MPa)
339	0	0	197	0	968	781	28	32.04
339	0	0	197	0	968	781	90	35.17
339	0	0	197	0	968	781	180	36.45
339	0	0	197	0	968	781	365	38.89
236	0	0	194	0	968	885	3	6.47
236	0	0	194	0	968	885	14	12.84
236	0	0	194	0	968	885	28	18.42
236	0	0	194	0	968	885	90	21.95
236	0	0	193	0	968	885	180	24.1
236	0	0	193	0	968	885	365	25.08
277	0	0	191	0	968	856	14	21.26
277	0	0	191	0	968	856	28	25.97
277	0	0	191	0	968	856	3	11.36
277	0	0	191	0	968	856	90	31.25
277	0	0	191	0	968	856	180	32.33
277	0	0	191	0	968	856	360	33.7
254	0	0	198	0	968	863	3	9.31
254	0	0	198	0	968	863	90	26.94
254	0	0	198	0	968	863	180	27.63
254	0	0	198	0	968	863	365	29.79
307	0	0	193	0	968	812	180	34.49
307	0	0	193	0	968	812	365	36.15
307	0	0	193	0	968	812	3	12.54
307	0	0	193	0	968	812	28	27.53
307	0	0	193	0	968	812	90	32.92
236	0	0	193	0	968	885	7	9.99
200	0	0	180	0	1125	845	7	7.84
200	0	0	180	0	1125	845	28	12.25
225	0	0	181	0	1113	833	7	11.17
225	0	0	181	0	1113	833	28	17.34
325	0	0	184	0	1063	783	7	17.54
325	0	0	184	0	1063	783	28	30.57
275	0	0	183	0	1088	808	7	14.2
275	0	0	183	0	1088	808	28	24.5

(continued)

(continued)

Cement (kg)	Blast furnace slag (mining) (kg)	Fly ash (kg)	Water (kg)	Water reducer (kg)	Coarse aggregate (kg)	Fine aggregate (kg)	Age (d)	Compressive strength (MPa)
300	0	0	184	0	1075	795	7	15.58
300	0	0	184	0	1075	795	28	26.85
375	0	0	186	0	1038	758	7	26.06
375	0	0	186	0	1038	758	28	38.21
400	0	0	187	0	1025	745	28	43.7
400	0	0	187	0	1025	745	7	30.14
250	0	0	182	0	1100	820	7	12.73
250	0	0	182	0	1100	820	28	20.87
350	0	0	186	0	1050	770	7	20.28
350	0	0	186	0	1050	770	28	34.29
203.5	305.3	0	203.5	0	963.4	630	7	19.54
250.2	166.8	0	203.5	0	977.6	694.1	90	47.71
157	236	0	192	0	935.4	781.2	90	43.38
141.3	212	0	203.5	0	971.8	748.5	28	29.89
166.8	250.2	0	203.5	0	975.6	692.6	3	6.9
122.6	183.9	0	203.5	0	958.2	800.1	90	33.19
183.9	122.6	0	203.5	0	959.2	800	3	4.9
102	153	0	192	0	887	942	3	4.57
102	153	0	192	0	887	942	90	25.46
122.6	183.9	0	203.5	0	958.2	800.1	28	24.29
166.8	250.2	0	203.5	0	975.6	692.6	28	33.95
200	133	0	192	0	965.4	806.2	3	11.41
108.3	162.4	0	203.5	0	938.2	849	28	20.59
305.3	203.5	0	203.5	0	965.4	631	7	25.89
108.3	162.4	0	203.5	0	938.2	849	90	29.23
116	173	0	192	0	909.8	891.9	90	31.02
141.3	212	0	203.5	0	971.8	748.5	7	10.39
157	236	0	192	0	935.4	781.2	28	33.66
133	200	0	192	0	927.4	839.2	28	27.87
250.2	166.8	0	203.5	0	977.6	694.1	7	19.35
173	116	0	192	0	946.8	856.8	7	11.39
192	288	0	192	0	929.8	716.1	3	12.79
192	288	0	192	0	929.8	716.1	28	39.32
153	102	0	192	0	888	943.1	3	4.78

(continued)

(continued)

Cement (kg)	Blast furnace slag (mining) (kg)	Fly ash (kg)	Water (kg)	Water reducer (kg)	Coarse aggregate (kg)	Fine aggregate (kg)	Age (d)	Compressive strength (MPa)
288	192	0	192	0	932	717.8	3	16.11
305.3	203.5	0	203.5	0	965.4	631	28	43.38
236	157	0	192	0	972.6	749.1	7	20.42
173	116	0	192	0	946.8	856.8	3	6.94
212	141.3	0	203.5	0	973.4	750	7	15.03
236	157	0	192	0	972.6	749.1	3	13.57
183.9	122.6	0	203.5	0	959.2	800	90	32.53
166.8	250.2	0	203.5	0	975.6	692.6	7	15.75
102	153	0	192	0	887	942	7	7.68
288	192	0	192	0	932	717.8	28	38.8
212	141.3	0	203.5	0	973.4	750	28	33
102	153	0	192	0	887	942	28	17.28
173	116	0	192	0	946.8	856.8	28	24.28
183.9	122.6	0	203.5	0	959.2	800	28	24.05
133	200	0	192	0	927.4	839.2	90	36.59
192	288	0	192	0	929.8	716.1	90	50.73
133	200	0	192	0	927.4	839.2	7	13.66
305.3	203.5	0	203.5	0	965.4	631	3	14.14
236	157	0	192	0	972.6	749.1	90	47.78
108.3	162.4	0	203.5	0	938.2	849	3	2.33
157	236	0	192	0	935.4	781.2	7	16.89
288	192	0	192	0	932	717.8	7	23.52
212	141.3	0	203.5	0	973.4	750	3	6.81
212	141.3	0	203.5	0	973.4	750	90	39.7
153	102	0	192	0	888	943.1	28	17.96
236	157	0	192	0	972.6	749.1	28	32.88
116	173	0	192	0	909.8	891.9	28	22.35
183.9	122.6	0	203.5	0	959.2	800	7	10.79
108.3	162.4	0	203.5	0	938.2	849	7	7.72
203.5	305.3	0	203.5	0	963.4	630	28	41.68
203.5	305.3	0	203.5	0	963.4	630	3	9.56
133	200	0	192	0	927.4	839.2	3	6.88
288	192	0	192	0	932	717.8	90	50.53
200	133	0	192	0	965.4	806.2	7	17.1

(continued)

(continued)

Cement (kg)	Blast furnace slag (mining) (kg)	Fly ash (kg)	Water (kg)	Water reducer (kg)	Coarse aggregate (kg)	Fine aggregate (kg)	Age (d)	Compressive strength (MPa)
200	133	0	192	0	965.4	806.2	28	30.44
250.2	166.8	0	203.5	0	977.6	694.1	3	9.73
122.6	183.9	0	203.5	0	958.2	800.1	3	3.32
153	102	0	192	0	888	943.1	90	26.32
200	133	0	192	0	965.4	806.2	90	43.25
116	173	0	192	0	909.8	891.9	3	6.28
173	116	0	192	0	946.8	856.8	90	32.1
250.2	166.8	0	203.5	0	977.6	694.1	28	36.96
305.3	203.5	0	203.5	0	965.4	631	90	54.6
192	288	0	192	0	929.8	716.1	7	21.48
157	236	0	192	0	935.4	781.2	3	9.69
153	102	0	192	0	888	943.1	7	8.37
141.3	212	0	203.5	0	971.8	748.5	90	39.66
116	173	0	192	0	909.8	891.9	7	10.09
141.3	212	0	203.5	0	971.8	748.5	3	4.83
122.6	183.9	0	203.5	0	958.2	800.1	7	10.35
166.8	250.2	0	203.5	0	975.6	692.6	90	43.57
203.5	305.3	0	203.5	0	963.4	630	90	51.86
310	0	0	192	0	1012	830	3	11.85
310	0	0	192	0	1012	830	7	17.24
310	0	0	192	0	1012	830	28	27.83
310	0	0	192	0	1012	830	90	35.76
310	0	0	192	0	1012	830	120	38.7
331	0	0	192	0	1025	821	3	14.31
331	0	0	192	0	1025	821	7	17.44
331	0	0	192	0	1025	821	28	31.74
331	0	0	192	0	1025	821	90	37.91
331	0	0	192	0	1025	821	120	39.38
349	0	0	192	0	1056	809	3	15.87
349	0	0	192	0	1056	809	7	9.01
349	0	0	192	0	1056	809	28	33.61
349	0	0	192	0	1056	809	90	40.66
349	0	0	192	0	1056	809	120	40.86
238	0	0	186	0	1119	789	7	12.05

(continued)

(continued)

Cement (kg)	Blast furnace slag (mining) (kg)	Fly ash (kg)	Water (kg)	Water reducer (kg)	Coarse aggregate (kg)	Fine aggregate (kg)	Age (d)	Compressive strength (MPa)
238	0	0	186	0	1119	789	28	17.54
296	0	0	186	0	1090	769	7	18.91
296	0	0	186	0	1090	769	28	25.18
297	0	0	186	0	1040	734	7	30.96
480	0	0	192	0	936	721	28	43.89
480	0	0	192	0	936	721	90	54.28
397	0	0	186	0	1040	734	28	36.94
281	0	0	186	0	1104	774	7	14.5
281	0	0	185	0	1104	774	28	22.44
500	0	0	200	0	1125	613	1	12.64
500	0	0	200	0	1125	613	3	26.06
500	0	0	200	0	1125	613	7	33.21
500	0	0	200	0	1125	613	14	36.94
500	0	0	200	0	1125	613	28	44.09
540	0	0	173	0	1125	613	7	52.61
540	0	0	173	0	1125	613	14	59.76
540	0	0	173	0	1125	613	28	67.31
540	0	0	173	0	1125	613	90	69.66
540	0	0	173	0	1125	613	180	71.62
540	0	0	173	0	1125	613	270	74.17
350	0	0	203	0	974	775	7	18.13
350	0	0	203	0	974	775	14	22.53
350	0	0	203	0	974	775	28	27.34
350	0	0	203	0	974	775	56	29.98
350	0	0	203	0	974	775	90	31.35
350	0	0	203	0	974	775	180	32.72
385	0	0	186	0	966	763	1	6.27
385	0	0	186	0	966	763	3	14.7
385	0	0	186	0	966	763	7	23.22
385	0	0	186	0	966	763	14	27.92
385	0	0	186	0	966	763	28	31.35
331	0	0	192	0	978	825	180	39
331	0	0	192	0	978	825	360	41.24
349	0	0	192	0	1047	806	3	14.99

(continued)

(continued)

Cement (kg)	Blast furnace slag (mining) (kg)	Fly ash (kg)	Water (kg)	Water reducer (kg)	Coarse aggregate (kg)	Fine aggregate (kg)	Age (d)	Compressive strength (MPa)
331	0	0	192	0	978	825	3	13.52
382	0	0	186	0	1047	739	7	24
382	0	0	186	0	1047	739	28	37.42
382	0	0	186	0	1111	784	7	11.47
281	0	0	186	0	1104	774	28	22.44
339	0	0	185	0	1069	754	7	21.16
339	0	0	185	0	1069	754	28	31.84
295	0	0	185	0	1069	769	7	14.8
295	0	0	185	0	1069	769	28	25.18
238	0	0	185	0	1118	789	28	17.54
296	0	0	192	0	1085	765	7	14.2
296	0	0	192	0	1085	765	28	21.65
296	0	0	192	0	1085	765	90	29.39
331	0	0	192	0	879	825	3	13.52
331	0	0	192	0	978	825	7	16.26
331	0	0	192	0	978	825	28	31.45
331	0	0	192	0	978	825	90	37.23
349	0	0	192	0	1047	806	7	18.13
349	0	0	192	0	1047	806	28	32.72
349	0	0	192	0	1047	806	90	39.49
349	0	0	192	0	1047	806	180	41.05
349	0	0	192	0	1047	806	360	42.13
302	0	0	203	0	974	817	14	18.13
302	0	0	203	0	974	817	180	26.74
525	0	0	189	0	1125	613	180	61.92
500	0	0	200	0	1125	613	90	47.22
500	0	0	200	0	1125	613	180	51.04
500	0	0	200	0	1125	613	270	55.16
540	0	0	173	0	1125	613	3	41.64
252	0	0	185	0	1111	784	7	13.71
252	0	0	185	0	1111	784	28	19.69
339	0	0	185	0	1060	754	28	31.65
393	0	0	192	0	940	758	3	19.11
393	0	0	192	0	940	758	28	39.58

(continued)

(continued)

Cement (kg)	Blast furnace slag (mining) (kg)	Fly ash (kg)	Water (kg)	Water reducer (kg)	Coarse aggregate (kg)	Fine aggregate (kg)	Age (d)	Compressive strength (MPa)
393	0	0	192	0	940	758	90	48.79
382	0	0	185	0	1047	739	7	24
382	0	0	185	0	1047	739	28	37.42
252	0	0	186	0	1111	784	7	11.47
252	0	0	185	0	1111	784	28	19.69
310	0	0	192	0	970	850	7	14.99
310	0	0	192	0	970	850	28	27.92
310	0	0	192	0	970	850	90	34.68
310	0	0	192	0	970	850	180	37.33
310	0	0	192	0	970	850	360	38.11
525	0	0	189	0	1125	613	3	33.8
525	0	0	189	0	1125	613	7	42.42
525	0	0	189	0	1125	613	14	48.4
525	0	0	189	0	1125	613	28	55.94
525	0	0	189	0	1125	613	90	58.78
525	0	0	189	0	1125	613	270	67.11
322	0	0	203	0	974	800	14	20.77
322	0	0	203	0	974	800	28	25.18
322	0	0	203	0	974	800	180	29.59
302	0	0	203	0	974	817	28	21.75
397	0	0	185	0	1040	734	28	39.09
480	0	0	192	0	936	721	3	24.39
522	0	0	146	0	896	896	7	50.51
522	0	0	146	0	896	896	28	74.99
273	105	82	210	9	904	680	28	37.17
162	190	148	179	19	838	741	28	33.76
154	144	112	220	10	923	658	28	16.5
147	115	89	202	9	860	829	28	19.99
152	178	139	168	18	944	695	28	36.35
310	143	111	168	22	914	651	28	33.69
144	0	175	158	18	943	844	28	15.42
304	140	0	214	6	895	722	28	33.42
374	0	0	190	7	1013	730	28	39.05
159	149	116	175	15	953	720	28	27.68

(continued)

(continued)

Cement (kg)	Blast furnace slag (mining) (kg)	Fly ash (kg)	Water (kg)	Water reducer (kg)	Coarse aggregate (kg)	Fine aggregate (kg)	Age (d)	Compressive strength (MPa)
153	239	0	200	6	1002	684	28	26.86
310	143	0	168	10	914	804	28	45.3
305	0	100	196	10	959	705	28	30.12
151	0	184	167	12	991	772	28	15.57
142	167	130	174	11	883	785	28	44.61
298	137	107	201	6	878	655	28	53.52
321	164	0	190	5	870	774	28	57.21
366	187	0	191	7	824	757	28	65.91
280	129	100	172	9	825	805	28	52.82
252	97	76	194	8	835	821	28	33.4
165	0	150	182	12	1023	729	28	18.03
156	243	0	180	11	1022	698	28	37.36
160	188	146	203	11	829	710	28	32.84
298	0	107	186	6	879	815	28	42.64
318	0	126	210	6	861	737	28	40.06
287	121	94	188	9	904	696	28	41.94
326	166	0	174	9	882	790	28	61.23
356	0	142	193	11	801	778	28	40.87
132	207	161	179	5	867	736	28	33.3
322	149	0	186	8	951	709	28	52.42
164	0	200	181	13	849	846	28	15.09
314	0	113	170	10	925	783	28	38.46
321	0	128	182	11	870	780	28	37.26
140	164	128	237	6	869	656	28	35.23
288	121	0	177	7	908	829	28	42.13
298	0	107	210	11	880	744	28	31.87
265	111	86	195	6	833	790	28	41.54
160	250	0	168	12	1049	688	28	39.45
166	260	0	183	13	859	827	28	37.91
276	116	90	180	9	870	768	28	44.28
322	0	116	196	10	818	813	28	31.18
149	139	109	193	6	892	780	28	23.69
159	187	0	176	11	990	789	28	32.76
261	100	78	201	9	864	761	28	32.4

(continued)

(continued)

Cement (kg)	Blast furnace slag (mining) (kg)	Fly ash (kg)	Water (kg)	Water reducer (kg)	Coarse aggregate (kg)	Fine aggregate (kg)	Age (d)	Compressive strength (MPa)
237	92	71	247	6	853	695	28	28.63
313	0	113	178	8	1002	689	28	36.8
155	183	0	193	9	1047	697	28	18.28
146	230	0	202	3	827	872	28	33.06
296	0	107	221	11	819	778	28	31.42
133	210	0	196	3	949	795	28	31.03
313	145	0	178	8	867	824	28	44.39
152	0	112	184	8	992	816	28	12.18
153	145	113	178	8	1002	689	28	25.56
140	133	103	200	7	916	753	28	36.44
149	236	0	176	13	847	893	28	32.96
300	0	120	212	10	878	728	28	23.84
153	145	113	178	8	867	824	28	26.23
148	0	137	158	16	1002	830	28	17.95
326	0	138	199	11	801	792	28	40.68
153	145	0	178	8	1000	822	28	19.01
262	111	86	195	5	895	733	28	33.72
158	0	195	220	11	898	713	28	8.54
151	0	185	167	16	1074	678	28	13.46
273	0	90	199	11	931	762	28	32.24
149	118	92	183	7	953	780	28	23.52
143	169	143	191	8	967	643	28	29.72
260	101	78	171	10	936	763	28	49.77
313	161	0	178	10	917	759	28	52.44
284	120	0	168	7	970	794	28	40.93
336	0	0	182	3	986	817	28	44.86
145	0	134	181	11	979	812	28	13.2
150	237	0	174	12	1069	675	28	37.43
144	170	133	192	8	814	805	28	29.87
331	170	0	195	8	811	802	28	56.61
155	0	143	193	9	1047	697	28	12.46
155	183	0	193	9	877	868	28	23.79
135	0	166	180	10	961	805	28	13.29
266	112	87	178	10	910	745	28	39.42

(continued)

(continued)

Cement (kg)	Blast furnace slag (mining) (kg)	Fly ash (kg)	Water (kg)	Water reducer (kg)	Coarse aggregate (kg)	Fine aggregate (kg)	Age (d)	Compressive strength (MPa)
314	145	113	179	8	869	690	28	46.23
313	145	0	127	8	1000	822	28	44.52
146	173	0	182	3	986	817	28	23.74
144	136	106	178	7	941	774	28	26.14
148	0	182	181	15	839	884	28	15.52
277	117	91	191	7	946	666	28	43.57
298	0	107	164	13	953	784	28	35.86
313	145	0	178	8	1002	689	28	41.05
155	184	143	194	9	880	699	28	28.99
289	134	0	195	6	924	760	28	46.24
148	175	0	171	2	1000	828	28	26.92
145	0	179	202	8	824	869	28	10.54
313	0	0	178	8	1000	822	28	25.1
136	162	126	172	10	923	764	28	29.07
155	0	143	193	9	877	868	28	9.74
255	99	77	189	6	919	749	28	33.8
162	207	172	216	10	822	638	28	39.84
136	196	98	199	6	847	783	28	26.97
164	163	128	197	8	961	641	28	27.23
162	214	164	202	10	820	680	28	30.65
157	214	152	200	9	819	704	28	33.05
149	153	194	192	8	935	623	28	24.58
135	105	193	196	6	965	643	28	21.91
159	209	161	201	7	848	669	28	30.88
144	15	195	176	6	1021	709	28	15.34
154	174	185	228	7	845	612	28	24.34
167	187	195	185	7	898	636	28	23.89
184	86	190	213	6	923	623	28	22.93
156	178	187	221	7	854	614	28	29.41
236.9	91.7	71.5	246.9	6	852.9	695.4	28	28.63
313.3	0	113	178.5	8	1001.9	688.7	28	36.8
154.8	183.4	0	193.3	9.1	1047.4	696.7	28	18.29
145.9	230.5	0	202.5	3.4	827	871.8	28	32.72
296	0	106.7	221.4	10.5	819.2	778.4	28	31.42

(continued)

(continued)

Cement (kg)	Blast furnace slag (mining) (kg)	Fly ash (kg)	Water (kg)	Water reducer (kg)	Coarse aggregate (kg)	Fine aggregate (kg)	Age (d)	Compressive strength (MPa)
133.1	210.2	0	195.7	3.1	949.4	795.3	28	28.94
313.3	145	0	178.5	8	867.2	824	28	40.93
151.6	0	111.9	184.4	7.9	992	815.9	28	12.18
153.1	145	113	178.5	8	1001.9	688.7	28	25.56
139.9	132.6	103.3	200.3	7.4	916	753.4	28	36.44
149.5	236	0	175.8	12.6	846.8	892.7	28	32.96
299.8	0	119.8	211.5	9.9	878.2	727.6	28	23.84
153.1	145	113	178.5	8	867.2	824	28	26.23
148.1	0	136.6	158.1	16.1	1001.8	830.1	28	17.96
326.5	0	137.9	199	10.8	801.1	792.5	28	38.63
152.7	144.7	0	178.1	8	999.7	822.2	28	19.01
261.9	110.5	86.1	195.4	5	895.2	732.6	28	33.72
158.4	0	194.9	219.7	11	897.7	712.9	28	8.54
150.7	0	185.3	166.7	15.6	1074.5	678	28	13.46
272.6	0	89.6	198.7	10.6	931.3	762.2	28	32.25
149	117.6	91.7	182.9	7.1	953.4	780.3	28	23.52
143	169.4	142.7	190.7	8.4	967.4	643.5	28	29.73
259.9	100.6	78.4	170.6	10.4	935.7	762.9	28	49.77
312.9	160.5	0	177.6	9.6	916.6	759.5	28	52.45
284	119.7	0	168.3	7.2	970.4	794.2	28	40.93
336.5	0	0	181.9	3.4	985.8	816.8	28	44.87
144.8	0	133.6	180.8	11.1	979.5	811.5	28	13.2
150	236.8	0	173.8	11.9	1069.3	674.8	28	37.43
143.7	170.2	132.6	191.6	8.5	814.1	805.3	28	29.87
330.5	169.6	0	194.9	8.1	811	802.3	28	56.62
154.8	0	142.8	193.3	9.1	1047.4	696.7	28	12.46
154.8	183.4	0	193.3	9.1	877.2	867.7	28	23.79
134.7	0	165.7	180.2	10	961	804.9	28	13.29
266.2	112.3	87.5	177.9	10.4	909.7	744.5	28	39.42
314	145.3	113.2	178.9	8	869.1	690.2	28	46.23
312.7	144.7	0	127.3	8	999.7	822.2	28	44.52
145.7	172.6	0	181.9	3.4	985.8	816.8	28	23.74
143.8	136.3	106.2	178.1	7.5	941.5	774.3	28	26.15
148.1	0	182.1	181.4	15	838.9	884.3	28	15.53

(continued)

(continued)

Cement (kg)	Blast furnace slag (mining) (kg)	Fly ash (kg)	Water (kg)	Water reducer (kg)	Coarse aggregate (kg)	Fine aggregate (kg)	Age (d)	Compressive strength (MPa)
145.4	0	178.9	201.7	7.8	824	868.7	28	10.54
312.7	0	0	178.1	8	999.7	822.2	28	25.1
136.4	161.6	125.8	171.6	10.4	922.6	764.4	28	29.07
154.8	0	142.8	193.3	9.1	877.2	867.7	28	9.74
255.3	98.8	77	188.6	6.5	919	749.3	28	33.8
272.8	105.1	81.8	209.7	9	904	679.7	28	37.17
162	190.1	148.1	178.8	18.8	838.1	741.4	28	33.76
153.6	144.2	112.3	220.1	10.1	923.2	657.9	28	16.5
146.5	114.6	89.3	201.9	8.8	860	829.5	28	19.99
151.8	178.1	138.7	167.5	18.3	944	694.6	28	36.35
309.9	142.8	111.2	167.8	22.1	913.9	651.2	28	38.22
143.6	0	174.9	158.4	17.9	942.7	844.5	28	15.42
303.6	139.9	0	213.5	6.2	895.5	722.5	28	33.42
374.3	0	0	190.2	6.7	1013.2	730.4	28	39.06
158.6	148.9	116	175.1	15	953.3	719.7	28	27.68
152.6	238.7	0	200	6.3	1001.8	683.9	28	26.86
310	142.8	0	167.9	10	914.3	804	28	45.3
304.8	0	99.6	196	9.8	959.4	705.2	28	30.12

The manufacturer's authorised representative in the EU is Springer Nature Customer Service Centre GmbH, Europaplatz 3, 69115 Heidelberg, Germany. If you have any concerns regarding our products, please contact ProductSafety@springernature.com

Printed and bound by CPI Group (UK) Ltd, Croydon, CR0 4YY

26/03/2026

02078974-0005